GROWTH, PHYSICAL ACTIVITY, AND MOTOR DEVELOPMENT

IN

PREPUBERTAL CHILDREN

GROWTH, PHYSICAL ACTIVITY, AND MOTOR DEVELOPMENT

— IN —

PREPUBERTAL CHILDREN

Toivo Jürimäe and Jaak Jürimäe

CRC Press
Boca Raton London New York Washington, D.C.

BS

Library of Congress Cataloging-in-Publication Data

Jürimäe, T.
    Growth, physical activity, and motor development in prepubertal children / Toivo
Jürimäe, Jaak Jürimäe.
        p. cm.
    Includes bibliographical references and index.
    ISBN 0-8493-0530-6
    1. Child development. 2. Motor ability in children. 3. Physical fitness in children. 4.
Children—Nutrition. 5. Pediatrics. I. Jürimäe, Jaak. II. Title.
    [DNLM: 1. Growth—Child. 2. Exercise—Child. 3. Motor Skills—Child. WS 103 J95 2001]
RJ47 .J87 2001
612.6'54—dc21
                                                   2001025108

Visit the CRC PressWeb site at www.crcpress.com

© 2000 by CRC Press LLC

No claim to original U.S. Government works
International Standard Book Number 0-8493-0530-6
Library of Congress Card Number 2001025108
Printed in the United States of America 1 2 3 4 5 6 7 8 9 0
Printed on acid-free paper

3-19-03

Dedication

To our father, Dr. Arnold Jürimäe

The Authors

Toivo Jürimäe, Professor and Ph.D., is Chair of the Department of Sport Pedagogy at the University of Tartu in Tartu, Estonia. Dr. Toivo Jürimäe graduated from the Faculty of Physical Education at the University of Tartu in 1973. He then pursued his doctorate in exercise physiology at the University of Tartu, graduating in 1980. He also completed an internship at the Charles University, Prague, Czech Republic, from 1982 to 1983. After working as a teacher of physical education in secondary schools from 1973 to 1977, he pursued his academic career at the University of Tartu, rising from Assistant Professor (1977) to Full Professor (1992).

Dr. Toivo Jürimäe is Vice President of the International Association of Sport Kinetics and a Board Member of the International Society for the Advancement of Kinanthropometry (ISAK) and Federation Internationale D´ Education Physique (FIEP). He is also a member of the American College of Sports Medicine, the European College of Sport Science, the European Anthropological Association, the International Council for Physical Activity and Fitness Research, and the Estonian Physical Education Association. He is an Editor of *Acta Kinesiologiae Universitatis Tartuensis,* a co-author of the monograph *Aerobic Exercises* (Moscow, 1988), and author and co-author of more than 150 research articles published in English, several in peer-reviewed international journals. He also has been an invited speaker at several international conferences.

Toivo Jürimäe's main research interests include body composition, physical activity, motor ability, and motor skills during growth and development. His research emphasizes health promotion and the reduction of coronary heart disease risk factors with physical activity.

Jaak Jürimäe, Ph.D., is a researcher at the Department of Sport Pedagogy at the University of Tartu in Tartu, Estonia. Dr. Jaak Jürimäe graduated from the Faculty of Physical Education at the University of Tartu in 1990. He then pursued his M.Sc. in exercise physiology at the University of Tartu, graduating in 1992. He pursued his Ph.D. at the University of Queensland (Australia), graduating in 1996. His entire academic career has been connected with the University of Tartu.

Dr. Jaak Jürimäe is an author and co-author of more than 40 research articles, most of them published in peer-reviewed international journals. He has frequently participated in international conferences. Dr. Jürimäe received The Young Investigator's Award at the Second Annual Congress of the European College of Sport Science in 1997. He is a member of the American College of Sports Medicine, the European College of Sport Science, the International Association of Sport Kinetics, and the Estonian Physical Education Association.

Jaak Jürimäe's main research interests include changes in muscle structure and function with different training loads, and growth and development in children. In addition, he has been interested in physical fitness and the performances of elite sportsmen. He emphasizes working with rowers.

Preface

Health of the adult population is closely interwoven with the health of children; and the health of children depends on their levels of physical activity, their motor abilities, and their motor skills. Children are born to move, to play, and to be physically active. The growth of children depends on the styles of living and nutritional habits of their parents. Active parents usually have active and physically skilled children.

Human growth and development have been extensively investigated and analyzed. Data are available about body stature and body mass of children in most countries, and data are also available about infancy and children older than 10 years of age.

Less attention has been given to studies of anthropometrical parameters, motor abilities, and motor development of children before puberty — ages 8 to 12 years. This is the period during which a child is at sexual maturation level 1 or 2, according to Tanner;[618] and chronological age may be different from biological age by 1 to 3 years. This period of somatic growth and development is very important since this is the time children begin their school careers — where the possibilities for voluntary play and movement rapidly decrease while mental stresses rapidly increase. This is also the time of so-called prepuberty; or, for some children (accelerants), this is the time for the beginning of pubertal changes. The beginning of puberty is particularly present at this age in children from southern countries. Children from northern countries tend to mature later.

Many methodologies exist to correctly measure physical activities, motor abilities, and motor skills of children. Several test batteries have been recommended for the measurement of motor abilities from agencies such as the American Alliance for Health, Physical Education, Recreation and Dance (AAHPERD), the Youth Fitness and Fitnessgram test batteries in the U.S., and the Eurofit test battery in Europe. However, universally accepted test batteries for the measurement of motor abilities in prepubertal children are not yet available.

Problems exist with correct, scientifically accepted measurements of physical activities in children of all ages. However, questionnaires and movement counters can give us some information. It is very important that young children have basic knowledge about correct running, jumping, throwing, and swimming as well as knowledge of how to play different sports and games. However, there are no criteria for acceptable levels of motor skills or how to correctly measure those motor skills.

The scientific measurement of body composition of prepubertal children is also a difficult task. Compared with adults, there are only a few recommendations of how to measure body composition in children using easy-to-perform field methods. Few regression equations are available for the calculation of fat mass and/or fat-free mass in prepubertal children, using a skinfold thickness or bioelectrical impedance analysis method. Furthermore, all regression equations depend on the group of children on whom the regression equation was validated. Other methods of body composition measurement in children appear to be time-consuming, expensive, or not suitable for children.

This book reports the results of comprehensive studies of development during the prepubertal years as they relate to environmental conditions, with special attention given to anthropometrical changes, physical activity, and motor development. Especially important are the longitudinal studies in which the growth of children is studied during several years or several decades (The Amsterdam Growth Study).[326] Of course, our knowledge today cannot answer all of the problems and questions related to somatic growth and development of prepubertal children.

This volume does not offer complete information on the rather broad topic of physical activity and fitness in prepubertal children, but it does cite certain examples from experimental studies. This book also includes information on the somatic growth and development of prepubertal children from Eastern Europe and the former Soviet Union countries — most of which has been unavailable for a broad audience. The material presented is not homogenous, but it does have one common denominator: physical activity is key to the growth and development of prepubertal children.

Contents

chapter one

Main factors influencing the development of prepubertal children

1.1 Biological maturation

The terms *growth* and *maturation* refer to distinct biological activities. Maturation is related to somatic, endocrinological, and psychological manifestations. Growth refers to measurable changes in body size, physique, and body composition — whereas biological maturation refers to progress toward the mature state.[57] Biological maturation varies not only among body systems but also in the timing of progress.[57] Growth focuses on size, and maturation focuses on the progress of attaining size.[57,201,407]

Chronological age is a poor marker of biological maturity in children. There are considerable varieties of physical characteristics among children of the same chronological age.[57,201,407] The processes of growth and biological maturation are related, and both influence physical performance in children.[57] According to Viru et al.,[656] the first critical period of biological maturation in regard to motor function is in infancy or early childhood. The second critical period appears at the age of 7 to 9, and the third critical period is during puberty.[656] Biological maturity of children can be estimated by different techniques that vary depending on the biological system being assessed. Most common systems include skeletal maturation, sexual maturation, and somatic (physique) maturation.[57,201,407,409]

Skeletal maturation is the best method to assess biological age or maturity status in children.[57,201,407] The skeleton develops from cartilage in the prenatal period to fully developed bone in early adulthood and serves as an easily identifiable indicator of biological maturation.[201,419,530] The assessment of skeletal age is based on the fact that a more mature child has more bone development and less cartilage than a less advanced child.[201,419,530] Bones of the extremities develop progressively during the growing years by ossification of cartilage.[419,530] Growth of these long bones occurs through proliferation of cartilage cells in the epiphyseal plate.[419,530] The mature adult state is reached when multiplication of these cells ceases and the bone fully ossifies.[201,419,530]

The x-ray of the hand–wrist area of the left hand is the site most often used to determine the amount of bone development and how near the shapes and contours of the bones are to adult status. The hand–wrist area of the left hand is ideal because it contains multiple bones for examination and is reasonably typical of the skeleton as a whole. Furthermore, the radiation exposure is minimal.

Three methods are available for assessing skeletal development of the hand and wrist — the Greulich-Pyle,[246] Tanner-Whitehouse,[619,620] and Fels[517] methods. All methods of skeletal maturity assessment are similar in principle and entail matching a hand–wrist x-ray of a child to a set of criteria — pictorial, verbal, or both. All three methods yield a skeletal age that corresponds to the level of skeletal maturity attained by a child in a reference sample.[201,409] However, these methods cannot be used interchangeably because they differ in scoring systems, they are based on different reference samples, and the skeletal ages derived are not equivalent.[407]

Sexual maturation is related to overall physiological maturation and can be used in estimating biological maturation. The assessment of sexual maturation is based on the evaluation of secondary sex characteristics such as breast development and menarche in girls, genital development in boys, and pubic hair development in both sexes.[201,409] However, the use of secondary sex characteristics is obviously limited to the pubertal phase of growth and maturation. Secondary sex characteristics are ordinarily categorized in five developmental stages for each characteristic as originally developed by Tanner.[618] Stage 1 indicates the prepubertal state, in which there is an absence of development of each secondary sex characteristic, while stage 5 represents the adult development.[618]

Another way to assess sexual maturation is to measure the level of sex hormones in the blood. A significant correlation has been found between the evaluation of sexual maturation pattern by sex hormones and by breast/genital developmental stages by Preece.[482] Growth hormone, secreted by the pituitary gland, is the major factor contributing to skeletal and somatic maturation during prepubertal years. The effects of growth hormone are mediated by somatomedins — substances produced by the liver that respond to increased growth hormone levels by stimulating tissue cell division and protein synthesis.[530] The function of growth hormone is supplemented by the effects of reproductive hormones at the onset of puberty.[530] In addition, depressed levels of thyroid hormone result in a delay in growth and sexual function.[530] Thus, it is obvious that the normal multiple functions of the endocrine systems are necessary for optimal biological maturation in prepubertal years.

Somatic maturation, manifested by a progressive increase in body size, is visually the most obvious expression of biological maturation of the child. However, the use of physique measurements as indicators of maturity status requires longitudinal data[201,409,530] to allow the timing estimation of the adolescent growth spurt. When adult stature data are available, the percentage of adult body size attained at different ages during growth can also be used as a maturity indicator.[409] Growth or growth rates between prepubertal boys and girls do not differ greatly.[201,409] In early adolescence, both stature and body mass accelerate in response to the hormonal changes of puberty.[530] Pubertal growth spurt occurs earlier in girls at about 10 to 12 years of age, with peak acceleration in boys appearing about two years later.[201,409,530] Peak height velocity is defined as the time during adolescence when maximum gains in stature are attained.[201,409] The age when the peak height velocity occurs is an indicator of somatic maturity.[201,409] Although ease and accuracy of measurement make somatic characteristics attractive as indicators of biological maturity, the use of stature and body mass to indicate the biological maturity of a child has major limitations.[201,409,530] First, stature and body mass demonstrate marked interindividual variabilities at given maturity levels. (For example, it is impossible to accurately determine the percentage of stature and body mass from the ultimate adult value in a 9-year-old boy). Second, the relationships between stature and body mass and other physiologic markers in a growing child are not simple.

The criteria for classifying children into maturity categories based on different maturation assessments have been suggested by Malina and Bouchard.[407] Using the skeletal age techniques, children assessed one year or less of their chronological age are classified as average maturers, while children whose skeletal age is delayed by more than one year are classified as late maturers. Those whose skeletal age is advanced by more than one year are classified as early maturers.[407] Similar categories have also been used with other maturation indicators, such as age of peak height velocity or age of menarche.[203]

Maturity-related somatic differences among children are most evident when comparing early maturers to late maturers. Malina and Bouchard[407] have noted that early maturers tend to be heavier and taller at all ages when they are compared to late maturers. However, final adult stature is generally similar. In terms of body shape, early maturers tend to be more endomorphic and mesomorphic, while late maturers tend to be more ectomorphic. Early maturers tend to have broader hips and narrower shoulders in comparison with late maturers, while late maturers tend to have greater leg lengths and shorter trunk lengths.[407]

Two main questions arise when using different maturity indicators to assess the biological maturation of a child.[409] First, do different maturity indicators measure the same kind of biological maturity? How closely in time do the curves of skeletal, sexual and somatic maturity indicators track together in a given child? Second, how consistent are maturity ratings over time? Is a child who is maturationally late at 5 years of age also late at 11 years of age? These questions could also apply to advanced and average maturity statuses.

Research data indicate that skeletal, sexual, and somatic maturity indicators are positively interrelated.[201,407,409,419,530] For example, correlation coefficients between such variables as peak height velocity, age of menarche, Tanner staging, and skeletal age are generally moderate to high, typically ranging from r = 0.60 to r = 0.80.[407,419,530] Thus a child classified as early, average, or delayed in maturity by one method will likely be classified similarly by other methods. Malina and Bouchard[407] have suggested that a general maturity factor underlies the maturation process during adolescence, placing children into early, average or late maturity groups. However, there is a variation within and among the maturity assessment techniques, which suggests that no single system will provide a complete description of an individual child's tempo of growth and maturation.[201,407]

Combining several skeletal, sexual, and somatic indicators into an overall maturity rating may be helpful in assessing the biological maturation of a child.[201,419] For example, McKay et al.[419] developed maturity ratings for prepubertal, early pubertal, and late pubertal children based on pubic hair self-assessment, height velocity (from four measurements over 2 years), menarcheal status for girls, and axillary hair rating for boys.

Skeletal maturity (bone age) is associated with somatic maturation (expressed as a percentage of adult stature) in the prepubertal years.[46,409] However, there is no clear relationship between bone age at onset of puberty and markers of sexual and morphological maturation later in adolescence.[201,413,419,530] This association becomes stronger as puberty progresses.[530] Furthermore, the tempos of prepubertal growth and maturation may be somewhat independent of pubertal growth and maturation.[409,530] For example, a cluster analysis of the Wroclaw Growth Study of Polish children, tracked longitudinally from 8 to 18 years of age, supports this idea.[61]

Two clusters were indicated by sexual maturity (ages at attaining genital and pubic hair; Tanner stages 2 and 4), skeletal maturity (at 11 to 15 years of age), and somatic maturity (ages at peak velocity for stature, body mass, leg length and trunk length, age at initiation of the stature spurt, and ages attaining 80, 90, 95, and 99% of adult stature).[61] The first cluster was a general maturity factor during adolescence. Clustering together were such measures as ages at peak velocities, attainment of stages of sexual maturation, skeletal ages at 14 and 15 years of age, ages at attaining 90, 95, and 99% of adult stature, and age at initiation of the stature spurt.[61] This general maturity factor suggests that the tempo of maturation during adolescence is under common control.[409,530] The second cluster identified clustering of skeletal ages at 11, 12, and 13 years of age as well as ages at attaining 80% of adult stature — both indicators of prepubertal growth and maturation. The second cluster was independent of the other maturity indicators.[61] This provided further evidence that prepubertal growth is independent of rates of biological development during adolescence.[409,530] This concept supports the notion that growth hormone is responsible for skeletal and somatic growth before puberty, while the influences of sex hormones become superimposed on the effects of growth hormone during puberty.[530]

The independence of prepubertal growth from biological development during adolescence raises the question of maturity indicators during childhood. Skeletal maturation is the primary indicator for prepubertal years since skeletal age is associated with body size.[46,409,419,530] The relationship between skeletal maturity and the attained percentage of adult stature is moderately high during the prepubertal years.[46,409] Children advanced in skeletal age are reportedly closer to adult stature at all ages before puberty and during adolescence when compared with those who are delayed in skeletal age relative to chronological age. The catch-up of those later in skeletal maturation occurs in late adolescence.[409] Variation in skeletal age is considerably reduced at menarche in girls and at peak height velocity in boys.[409] For example, mean chronological age and skeletal age at menarche in girls are 13.2 and 13.3 years, respectively,[413] while in boys, Malina and Beunen[409] have suggested a high correlation (r = 0.81) between skeletal age at 14 years and age at peak height velocity (13.9 years). Thus, the correlations between maturational events that occur closer in time are higher than those separated in time.[409]

In general, prepubertal growth appears to be independent of the biological maturation during puberty. Biological maturation during prepubertal years is best determined by skeletal age. However, skeletal age cannot be used alone as a method to predict biological maturation when pubertal events occur. As adolescence progresses, skeletal maturity is increasingly related to the indices of sexual and somatic maturation. The mechanisms that determine biological maturation during growth are complex and not easily explained. However, the examination of factors that influence the rates of skeletal, sexual, and somatic maturation is important to understanding the complexity of growth. Factors such as genetics, nutrition, endocrine, physical activity and inactivity, and social conditions can influence the process of biological maturation in children. Furthermore, differences in race, culture, climate, and geographic location can also affect these changes. Finally, childhood illness may also affect normal rates of growth and maturation and cause deviations from expected curves.[530]

Genetics significantly influence growth.[419,530] Children inherit genetic information from their parents and resemble their mothers or fathers in size and physical characteristics. Studies of twins have suggested that the genetic contribution to a child's stature as well as to his or her eventual adult stature is approximately 60%. The correlation coefficient between a child's stature at 3 years of age and that at maturity is r = 0.80.[419] The effect of heredity factors on body mass has been reported to be lower at about 40%.[407] The genetic contribution to length and diameter of long bones appears to be about 60%.[419] The correlation coefficients for age at menarche between monozygotic and dizygotic twins have been found to be r = 0.90 and r = 0.60, respectively.[206]

The most critical environmental factor influencing normal biological maturation is appropriate nutrition. An inadequate caloric intake or deficiency in any specific dietary component can impair normal biological maturation. Appropriate food intake parallels increase in body size, since the child has to consume sufficient food to provide enough calories for normal growth. Furthermore, the diet must be composed of an appropriate balance of protein, fat, carbohydrates, vitamins, and minerals to support growth.[530]

The level of habitual physical activity has no effect on body stature in children. However, increased physical activity has a positive effect on body mass, diminishing the level of body fat and increasing the level of muscle mass. Furthermore, prolonged training may result in increased bone density. However, physical activity does not affect the rate of skeletal maturation.[530]

Finally, chronological age is seldom equal to the level of skeletal, sexual, or somatic maturation in children. It is important to know the individual level of biological maturation of children, characterizing, for example, the level of motor abilities. Early maturers have an advantage in motor abilities during prepubertal years. However, their development levels will be the same or even lower during their later years in comparison with children with late biological maturation. We know several examples of high-level athletes who were late maturers and whose levels of motor abilities and anthropometrical parameters were lower than average during prepubertal years. Several simple methods (Tanner stages or anthropometry) and relatively complicated and expensive methods (x-ray or hormonal analyses) are recommended for the measurement of biological maturation. However, it is best to use more than one parameter to reach a final conclusion.

1.2   Influence of environmental factors

Several factors such as climatic zones, traditions, or availability of sports facilities can influence the development of children. Sports events are different for children of different countries. For example, in the Nordic countries such as Finland and Sweden, very young boys begin to play ice hockey or they study the elementary skills of cross-country skiing. In contrast, in South-American countries such as Brazil and Argentina, soccer is the first choice of prepubertal boys. There are also differences between urban and rural areas in demographics and socioeconomic factors that contribute to differences in physical activity patterns. For example, in Finland a rural environment still exists; and the differences between urban and rural environments influence the physical activities of children. Urban children have more options for physical activities, more information about sports, and more opportunities to utilize sports facilities and equipment than rural children have. People living in small villages are separated by long distances, making it difficult to organize group sports activities.[357] There is usually a lack of sports facilities in small villages. On the other hand, the countryside offers better conditions for outside physical activities such as roaming the forest or swimming in lakes.

There is an increased focus on determining the relationship between physical activity and such psychosocial correlates as family and peer support, intrinsic motivation, and confidence and self-esteem to gain a better understanding of physical activity behavior in children.[635]

Family members significantly impact the physical activity pattern of children. More than 20 years ago, Greendorfer and Lewko[245] suggested that fathers appeared to be more important socializing agents than mothers during childhood. But it has also been suggested that: (1) parental physical activity patterns, as a rule, influence only the physical activity patterns of daughters;[126] (2) mothers' physical activity patterns influence the physical activity levels of children;[421] and (3) parental physical activity patterns primarily influence children of the same sex (fathers have more influence on sons, and mothers have more influence on daughters).[683] The longitudinal study of Finnish children indicated the critical influence of fathers on the physical activity levels of children.[688] Thus, it can be concluded that parental roles, especially fathers' roles, are important in increasing the levels of physical activities of children. Interestingly, however, Freedson and Evenson[220] concluded that the inactivity of parents may exert more influential modeling behavior than physical activity.

In a family, as a rule, boys and girls are treated differently.[364] Boys are allowed more freedom to display aggressive behaviors and to engage in more vigorous physical activities, whereas girls are encouraged to be more dependent and less exploratory in their behaviors. Mothers and fathers elicit gross motor behaviors more from their sons than their daughters. The influence of peers on the physical activity levels of prepubertal children is poorly studied. Only Stucky-Ropp and Di Lorenzo[611] have indicated that the support of peers is more pronounced in 11-year-old children than in younger children.

Self-perception is one of the most important concepts in understanding human behavior. The self is defined in relation to normal physical and social standards during the prepubertal ages of 8 to 12 years. At this stage, children compare their own performances and capabilities with those of others, real or imaged.[144] During this period, the lives of children expand outside their homes. Organized physical education lessons and organized sports begin to affect the development of children. However, children are not able to understand that abilities, capacities, effort, and practice all affect the final result during the prepubertal time. In connection with this, it is not recommended to overemphasise and overorganize competitions for children because children do not understand comparison and competition in the same manner that adults do.[370] On the other hand, children clearly make distinctions between competence domains in the cognitive, social, and physical realms beginning, on average, at the age of 8.[370]

The most important environmental factor for prepubertal children is the ability to play outside 12 months a year. Of course, the amount of outdoor activities depends on the season, geographical location, etc. Unfortunately, in today's world children cannot be outside as much as children could a few decades ago. Armstrong and Bray[16] indicated that there are no differences in the levels of physical activity in 10- and 11-year-old children in summer and autumn terms. However, the outdoor activities likely peak in summer, fall off in autumn, diminish further in winter, and then rise again during spring. It is very important to use summer holidays to increase the levels of physical activity of children. It is important to have access to indoor swimming pools, sports halls, etc., during the colder seasons. Finally, 84.4% of the average weekly minutes of participation in physical activities occurs outside school physical education, during recess, in other unstructured play, or with structured park and recreation programs, community sports teams, or religious groups.[588] For example, mile walk/run test results are well correlated with the total amount of out-of-school activities in children.[526] Hillman's[281] research has clearly demonstrated that outside activities of children have declined with time. Furthermore, boys enjoy far more independence than girls in each of the situations studied.[281] For example, one third of boys owning bicycles were allowed to cycle on the roads, whereas only one in nine of girls owning bicycles were allowed to do so in England.[281]

Environmental factors strongly influence the development of prepubertal children either directly or indirectly. Parents control access to environments that facilitate physical activity, such as playgrounds, and they influence sports participation by transporting children to sports facilities. Furthermore, they determine the frequency and length of time children spend outdoors. Outdoor activities also depend on prevailing climatic conditions, which influence what children can do outdoors (winter or summer activities). Children need some elementary knowledge about different physical activities before puberty.

1.3   Health, physical activity, and inactivity

Physical activity in different forms is important to health and development during childhood.[31] Children need regular physical activity for normal growth and development, maintenance of good health and fitness, and development of physical activity skills and behaviors that carry into adulthood.[588] All countries want to increase the levels of physical activity in children, and these increases should be national priorities.[488] However, there is a widespread decline of school physical education in most European countries and an associated perception that children´s freedom to play, walk, or cycle outdoors is restricted. Further, children spend too much time watching television, playing video games, and generally adopting a lifestyle of sedentary pursuits. However, very few children are so unfit that their functional capabilities for daily living are impaired.

Establishing the association between habitual physical activity and health outcomes is more difficult in children than in adults because:[556]

1. The school year restricts physical activity.
2. Disease risk and health behaviors have less variation because environmental influences have not had as much time to exert their effects.
3. The effects of exercise on health or disease risk may not have had enough time to become evident.

The level of physical activity in children has declined during recent years. However, the newest investigations in Finland indicate that children are very active during their leisure time at prepubertal ages.[554] Physical activities of boys and girls appear to be similar in summer, while boys exercise more intensively during winter. In one of the pioneer studies on the influence of physical education lessons during a one-year period, Cumming et al.[139] reported no impact on cardiorespiratory fitness (physical working capacity [PWC_{170}] and maximal oxygen consumption [VO_{2max}]), regardless of the number of hours of physical education during each week. In contrast, Shephard and Lavallee[578] indicated a *significant* impact of increased physical education hours on PWC_{170} and VO_{2max} in prepubertal children. In some cases, physical education lessons can improve running test performances in first-grade children.[395] A year-long Portuguese study of 9-year-old children[385] indicated that physical education lessons had no effect upon cardiorespiratory endurance, flexibility, and body fat. The lessons had a positive impact upon sit-ups in 60 seconds and modified pull-up performance. The study concluded that children with three physical education lessons a week and children following the alternative program oriented primarily to sports (soccer, basketball, handball, gymnastics, and track and field) had greater improvements than the children with two physical education lessons a week and following the official school curriculum.[385]

The efficiency of the physical education lessons also depends on the qualification of the teachers. Physical education should be taught by a certified physical education teacher. In elementary school, as a minimum, it should be taught by a classroom teacher who has received special training in physical education. Most of the time should be devoted to moderate to vigorous physical activity in high-quality lessons that are held in comfortable sports facilities.

Compulsory physical education at school does not compensate for the lack of physical activity in children. In Europe, the time devoted to physical education in youths between the ages of 6 to 18 years varied by country from 30 minutes a week (Ireland) to 150 minutes a week (France) — far less than needed.[259] Current levels of physical activity must be enhanced for prepubertal children by increasing the number of obligatory physical education lessons to at least one lesson per school day. The quality of the lessons must also be elevated with more emphasis on improving physical skill development and on motivating children to be physically active in their free time.

An important function of school physical education is to engage children in moderate to vigorous physical activity, a requisite for health and motor skill development.[588] Cardiorespiratory fitness of children is only promoted when children spend an appropriate amount of time in moderate to vigorous physical activity (3 lessons a week) — which most curricula do not offer.[588] Many physical education programs are available, and the dominant curricula emphasizes skill-related fitness of movement — while developmental, humanistic and personal curricula have been promoted.[344] During the past 10 to 15 years, as the important health effects of physical education have been documented, the emphasis has shifted to the effects of physical education on health-related physical activity and fitness.[66,488] Several studies have demonstrated that health-related physical education programs increase physical activity during class and improve cardiovascular fitness in elementary school.[420]

Sallis et al.[553] evaluated the effects of a health-related physical education program on fourth- and fifth-grade students. Sallis et al.[553] concluded that health-related physical education programs provide students with substantially more physical activity. In addition, Shephard and Lavallee[579] suggested that enhanced physical education programs (one hour of required physical education daily) improved several indices of physical performance in elementary schoolchildren. Simons-Morton et al.[591] revealed that the modification of the school physical education program increased the time children engaged in moderate to vigorous physical activity from less than 10% to about 40%.

The effect of physical education lessons on body fat content is usually not significant.[395,577] However, Johnson and La Von[307] indicated positive effects of physical education upon body fat. Measurable benefits of regular childhood physical activity include improved cardiorespiratory fitness, strength, weight control, and body composition.[599] In childhood, linkages between physical activity and early disease indicators such as obesity[393] and high blood pressure[81] have been identified. There is an increasing concern that childhood obesity and the development of cardiovascular disease risk factors are directly related to a child's sedentary living. Cardiovascular disease is now recognized as a pediatric problem. An inverse relationship exists between physical fitness and cardiovascular risk factors in children. Physical fitness is the variable with the highest and most consistent impact on cardiovascular disease risk factors — providing an influence greater than that of obesity. An increased polarization of physical fitness is mirrored in the fact that obese children have become even more obese, and they have increased in number.

The exact mechanism of the protective effect of physical activity has not been clearly established, partially because the level of physical activity of children is highly variable. Different determinants of physical activity should be considered, including intensity, time of activity, metabolic efficiency, overall energy cost, and type of physical activity (school obligatory physical education lessons, recreational, spontaneous movement, different games, etc.). Aspects of different physical activities need to be considered, such as quantitative (energy cost), qualitative (type and duration of activity), and the effects of physical activity on an intermediary metabolism.[211] Surprisingly, Goran et al.[239] demonstrated that activity-related energy expenditure was not related to body fat, gender, or parental weight status in young children. In another study, the same group of researchers[241] indicated that energy expenditure was highly variable and only weakly related to fat-free mass and body mass in young children. In addition, the daily energy cost of physical activity was unrelated to time of day. However, Goran et al.[241] demonstrated that body fat mass was more related to activity time than to the combined energy cost of physical activities.

One of the most powerful health determinants, body fat, demonstrates a negative relationship with physical activity.[35,36] Dietz and Gortmaker[164] indicate that increased television watching is the major risk factor for the development of obesity in children. In contrast, Lindquist et al.[369] demonstrate that children at the ages of 6.5 to 13 years who watch more television do not necessarily engage in less physical activity.

The physical activities of children are very complex and depend on several factors. To study the development of prepubertal children, complex methodology is recommended that includes biological and sociocultural determinants. One of the classifications was recently presented by Kohl and Hobbs[342] that defines the determinants of development at four levels:

- Physiological — maturation, growth
- Psychological — motivation, self-efficacy, sense of control
- Sociocultural — family characteristics, sociodemographics, role models
- Ecological — facilities, physical safety, climate

Most prior research has concentrated on physiological factors, and several excellent studies have also focused on the psychological factors.[635] Broader sociocultural[369] and ecological[342] determinants have received little attention, despite their potential to influence activity patterns of children. Krombholz[351] reported an inverse relationship between socioeconomic background and physical activity and fitness in children in kindergarten and elementary school. The level of physical activity in children depends also on the family size. Compared to children residing in two-parent homes, children with single parents report more hours of watching television, less exercise received in school physical education classes, and less days per week of exercise.[369]

In comparison to adults, children have had fewer investigations into coronary heart disease (CHD) risk factors and their relationship to physical activity and physical fitness. Malina[401] indicated that relationships among physical activity, fitness, and CHD risk factors are influenced by the heterogeneity in biological maturation. There are contradictory data about the influence of physical activity on CHD risk factors. In the Cardiovascular Risk in Young Finns Study,[492] active males (9 to 24 years old) had lower triglycerides and higher high-density lipoprotein concentrations (both HDL and HDL_2) than inactive males. Active females had lower triglyceride levels than inactive females when controlling for pubertal status. In contrast, the Singapore Youth Coronary Risk and Physical Activity Study[562] demonstrated that the relationship between CHD risk factors and physical activity was low. Interestingly, as early as 3 to 4 years of age, physical activity is related to several CHD risk factors according to Sääkslahti et al.[537]

The results between CHD risk factors and physical fitness are also contradictory. For example, Gutin et al.[249] reported a significant relationship between VO_{2max} and atherogenic index in a sample of boys and girls 7 to 11 years old. In contrast, Suter and Hawes[616] reported a nonsignificant relationship between VO_{2max} and HDL in boys and girls. A significant correlation between VO_{2max} and HDL in children 5 to 12 years old has also been reported.[434] However, after adjustment for age, sex, fat, and triglycerides, this relationship lost significance.[434] Kwee and Wilmore[356] found no differences in HDL among boys 8 to 15 years old classified into four fitness categories based on directly measured VO_{2max} levels. Using the multivariate design of study, Katzmarzyk et al.[323] indicated significant relationships among physical activity, physical fitness (submaximal physical working capacity on a bicycle ergometer [PWC_{150}]), and CHD risk factors in youths 9 to 18 years old. Physical fitness explained a slightly greater proportion of the variance in CHD risk than did physical activity. Results of the Oslo Youth Study[624] indicated that there was an inverse relationship between physical fitness (indirectly assessed VO_{2max} on a bicycle ergometer) and CHD risk factors in Norwegian children 10 to 15 years old. Thus, it can be concluded that the relationships among physical activity, physical fitness, and different CHD risk factors in children are quite complex; and results to date are inconclusive.

There are numerous studies about tracking CHD risk factors in children. For example, children who have high blood pressure tend to become hypertensive adults. It has been reported that childhood blood pressure significantly predicts blood pressure levels 15 years later.[686] The concentration of blood lipids has much higher tracking coefficients, and tracking has been documented up to 9 consecutive years.[459] This indicates that changing risk factors in childhood may influence CHD risk factors in adulthood. It is difficult to expect high tracking correlations of these variables since they respond to different factors — genetic, environmental, and behavioral — and there are complex interactions among these variables as well. Longitudinal studies are warranted to better understand the effects of childhood physical activity on the tracking of CHD risk factors.

Optimal levels of physical activities are powerful agents for decreasing the risk factors of different diseases that can guarantee good health of prepubertal children. However, it is not easy to define the optimal level of physical activity for every child since it is different for every child. Thus, only general recommendations can be given for prepubertal children.

chapter two

Anthropometric development of prepubertal children

2.1 Introduction

The main purpose of anthropometry is to assess and monitor growth. Growth in stature and body mass are frequently used as markers of health and nutritional status of children.[466,520,521,528,582] More detailed data on growth, including further anthropometric measurements such as length, breadth, circumference, and skinfold variables[381,448,528] are not well documented. Similarly, the use of body composition parameters to monitor the growth patterns of children is less common. Anthropometric tracking of these measures, together with motor ability and skills values, provides more information on the developmental process of children.

The growth pattern of a child is the result of a continuous interaction between the child's genes and environment. This includes the socioeconomic environment of the family and school as well as the ecological environment of the district and country. Changes in the growth pattern, therefore, reflect changes in one or more of these factors.[368] It is not easy to assess the extent to which different variables such as genetics, growth hormones, maturity timing and rates, nutrition, and physical activity affect the anthropometric development of children. All variables are important in physique changes. However, length and breadth measures of the skeleton are more genetically determined than body mass and skinfold thicknesses, which are more environmentally dependent.[109]

2.2   Somatic growth

Body size and proportions, physique, and body composition are important factors in growth and anthropometric development of prepubertal children. Historically, body stature and mass, both indicators of overall body size, have extensively been used with age and sex to identify the anthropometric development of children. Body size, particularly body mass, is a standard frame of reference for expressing physiological parameters in children. Physique is the body form of an individual — the configuration of the entire body rather than of specific features — commonly referred to as body build. Physique is readily observed and is useful in assessing the outcomes of underlying growth and maturation processes, thus leading to a better understanding of variation in both child and adult physiques.

Organs grow at different rates, and these rates can differ from the growth rate of the human body as a whole.[200,520,521,618] Furthermore, children can grow up in a normal process, where growth is organized in successive steps, or the growth process may be influenced by an individual variation due to genetic and/or environmental factors.[521] This variation makes it difficult to predict adult body composition from childhood measurements. However, numerous growth grids from children of different countries have been prepared and used for the evaluation of growth levels in children.[73,466,481,521,523,653] Most research addresses the whole growth period from birth to maturity, especially those focused on ontogenetical changes in stature and body mass.[73,466,481,520,521,523,653]

With the specific exception of the sex organs, there are only minor differences in anthropometric characteristics between boys and girls through to the age of puberty. Before puberty, boys and girls have similar average statures and body masses.[582] However, girls tend to be a little fatter than boys from an early age.[114] Girls commence their pubertal growth spurt one or two years earlier than boys, at around 10 years of age. For a short period, girls are taller than boys of similar age. The later pubescent spurt in boys allows growth to continue about two years longer than in girls. This delay in boys is responsible for their greater adult stature and their longer legs and arms relative to stature.[114] In males, the dominant change in puberty is an increase of muscle, although there is also some increase of subcutaneous fat over the abdomen and chest. Pubertal maturation status is based on the development of breasts and pubic hair in girls and pubic hair and genitals in boys.[520,618]

Continued growth of a child is generally considered a sign of health and well-being. Linear growth velocity decreases rapidly from 30 cm a year during the first months of life, to approximately 9 cm a year at the age of 2, and to 7 cm a year at 5 years of age. Linear growth rate then continues at approximately 5.5 cm a year before slowing slightly just before puberty.[466,520,521] For girls, who follow a typical growth curve, the pubertal growth spurt begins at approximately 10 years of age, reaches a peak of approximately 10.5 cm a year at the age of 12, and then decreases toward zero around the age of 15.[520,521] For an average boy, the growth velocity increases sharply around the age of 12, reaches a peak velocity of 12 cm a year at the age of 14, and then decelerates toward zero around the age of 17.[520,521]

Body mass velocity decreases sharply from approximately 10 kg a year during the first 2 years, and then it accelerates slowly throughout the remainder of childhood to 3 kg a year in both sexes. During puberty, girls attain a peak body mass velocity of 8.5 kg a year at approximately age 13, and boys attain 9.5 kg a year at approximately age 14. In both sexes, the peak body mass velocity is followed by a quick decrease to less than 1 kg a year for girls at age 15 and for boys at age 17.[520,521]

Recent research[313,454,455,495,652] at the University of Tartu on cross-sectional groups of boys and girls from 4 to 17 years of age has led to the development of growth grids for stature and body mass for Estonian children (Figure 2.1). The growth grids for stature and body mass of Estonian children are comparable to the World Health Organization (WHO) reference population.[685] Accordingly, the conditions of growth and development for Estonian boys and girls are comparable to those in developed countries. The average stature

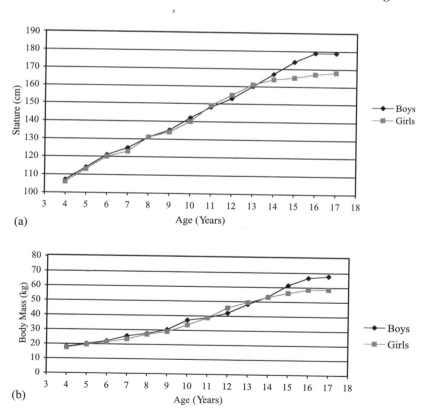

Figure 2.1 Comparison of average values of stature (a) and body mass (b) in Estonian boys and girls aged 4 to 17 years.

of Estonian boys and girls increases approximately 5.5 cm a year through to the age of puberty, while the average body mass of Estonian children increases approximately 3 kg a year up to the age of 11 (Figure 2.1). The same growth pattern for stature and body mass was also observed for Estonian girls measured in the mid 1980s.[626]

Anthropometric measurements can be used in several ways to study the growth of children:[521]

- Directly (skinfolds, circumferences, breadths, diameters)
- As indices (body mass to stature squared, the body mass index [BMI])
- Areas (upper arm muscle area based on arm skinfolds and arm circumference)
- Regression equations relating body density to anthropometric measurements for a reference population

In addition, various ratios can be used to predict body shape and proportion during growth. However, more detailed data on growth using these different anthropometric parameters are less common.[466] Specific anthropometric measurements used in children during growth will be discussed in Chapter 2.3.

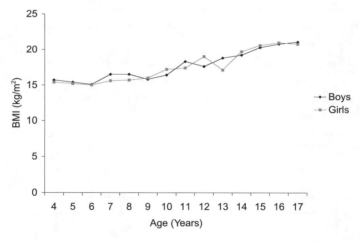

Figure 2.2 Body Mass Index (BMI) (body mass $[kg]$/stature$^2[m^2]$) variation in Estonian children aged 4 to 17 years.

Growth curves for BMI have been developed for children in France,[521,522] the U.K.,[523] the Czech Republic,[486] the U.S.,[515,582] the Netherlands,[524] and in other countries.[466,521,614] Growth curves for BMI are also presented for Estonian children (Figure 2.2).[313,454,455,495,652] The longitudinal evolution of BMI has also been investigated.[523,582] The BMI has been found to be associated with body composition and nutritional status.[466,515,522,614] It also has a high correlation with total body fat,[316,448,466,515,521,614] more specifically by the subscapular skinfold thickness,[432,448,614] and a low correlation with stature.[521] During growth, body mass increases with both age and stature, and these associations reflect changes in stature rather than changes in body fat.[466,521] However, the last research by Siervogel et al.[583] illustrates the complexity of interpreting changes in BMI in individual children. Furthermore, they reinforce the fact that the BMI is a measure of body mass, not adiposity per se.[583] The usefulness of BMI in children is complicated by its dependency on stature, relative differences among trunk and leg length, fat-free mass, and maturity level. Siervogel et al.[583] concluded that BMI is a useful tool in helping to define overweight and obesity in children.

Differences in socioeconomic conditions should also be taken into account when characterizing the anthropometric parameters of prepubertal children.[466,614] For example, studies in Sweden,[368] Norway,[95] and Hungary[72] have demonstrated that, concomitant with a disappearance of social differences in stature, the children of the lower social classes were heavier for the same stature. A greater body mass for stature and a greater prevalence of obesity in children still seem to characterize lower socioeconomic groups.[72,614] This tendency starts early in life and is probably linked to dietary differences.[614] High intake of protein during childhood is considered one of the main causes of later obesity.[521] Thus, the higher values of BMIs reflect a high adiposity level; and the origins lie in a combination of bad eating habits and lifestyles from early childhood.[196,466,521,583,614]

A comparison of the values of BMI in Estonian children (Figure 2.2) with the children of other countries demonstrates that Estonian children are similar to the children of the Czech Republic.[486] However, Estonian children tend to be heavier than French children.[521,522] Regarding health, research has demonstrated that there are more overweight people in Eastern European countries compared with most industrially developed countries of Western Europe, the U.S., and Japan.[642] However, the prevalence of higher BMI values has also increased considerably in the U.S.[679] Socioeconomic conditions in Estonia are still subject to changes that tend to bring about growing diversity within the population and are likely to have both positive and negative effects on the growth of Estonian children.

2.3   Main anthropometric parameters

Anthropometry involves the measurement of carefully defined body landmarks to provide information on the size of the individual as a whole (stature and body mass) and of specific segments, parts, and tissues. Skeletal breadths describe the overall robustness of the skeleton, limb circumferences provide information on relative muscularity, and skinfold thicknesses indicate subcutaneous fat. The specific dimensions include both the trunk and the extremities, because children can be similar in overall body size but vary in shape, proportion, and tissue distribution during growth.

A recent cross-sectional study to investigate the anthropometric profile of Estonian prepubertal children was carried out according to the protocol recommended by the International Society for the Advancement of Kinanthropometry.[448] In total, the following measurements were made:[316]

- 9 skinfolds — triceps, subscapular, biceps, iliac crest, supraspinale, abdominal, front thigh, medial calf, and mid-axilla
- 13 circumferences — head, neck, arm relaxed, arm flexed and tensed, forearm, wrist, chest, waist, gluteal, thigh, thigh mid trochanter-tibiale laterale, calf, and ankle
- 8 lengths — acromiale-radiale, radiale-stylion, midstylion-dactylion, iliospinale-box height, trochanterion-box height, trochanterion-tibiale laterale, tibiale-laterale to floor, and tibiale mediale-sphyrion tibiale
- 8 breadths/lengths — biacromial, biiliocristal, foot length, sitting height, transverse chest, A-P chest depth, humerus, and femur

The skinfold thicknesses were measured using Holtain (Crymmych, U.K.) skinfold calipers, while all other anthropometric measurements were made using Centurion Kit instrumentation (Rosscraft, Surrey, B.C., Canada). Because detailed data on anthropometric development of 9- to 11-year-old prepubertal children from the Baltic region of Eastern Europe are relatively rare, all measured variables are given in Table 2.1. With sex-linked differences, the skinfold thicknesses measured on the triceps, subscapular, biceps, and mid-axilla sites — and the sum of all measured skinfolds — were significantly higher in prepubertal girls.[316] Thus, the subcutaneous fat appears to be more pronounced in prepubertal girls. This is in accordance with the results of other investigations on prepubertal children.[154,204,303,309,597,634]

Most of the measured circumferences were significantly higher in boys.[316] The only circumferences that were not different between sexes were gluteal, thigh, and calf circumferences, suggesting that the relative muscularity of the lower body is developed in a similar manner in prepubertal boys and girls. Thus, sex-linked differences in circumferences on the upper body region were somewhat more marked than those of the lower body region. This pattern has been observed in preschool children.[204,466]

The overall robustness of the skeleton was determined according to the selected length and breadth/length parameters of the upper and lower body (Table 2.1). Also with sex-linked differences, the comparison of measured length parameters revealed higher values in boys, although these differences did not reach statistical significance in all cases.[316] The dimensions of the upper and lower extremities were significantly higher in boys. Sex-linked differences in breadth/length parameters were applied also on the trunk region, with the values significantly higher in prepubertal boys.[316] Thus, the robusticity of skeleton is more developed in prepubertal boys in comparison with girls of the same age.

Table 2.1 Mean Anthropometric Variables of
Estonian Prepubertal Children (X ± SD)

Measured Variables	Boys (n = 104)	Girls (n = 105)
	Profile	
Age (yrs)	10.1 ± 0.8	9.8 ± 0.7[a]
Stature (cm)	143.4 ± 7.3	141.5 ± 7.3
Body mass (kg)	35.3 ± 5.7	33.3 ± 6.4[a]
	Skinfolds (mm)	
Triceps	10.0 ± 3.0	11.2 ± 3.9[a]
Subscapular	7.3 ± 3.5	8.4 ± 5.1[a]
Biceps	6.6 ± 2.5	7.5 ± 3.4[a]
Iliac crest	8.7 ± 4.8	9.3 ± 5.5
Supraspinale	5.1 ± 2.6	6.2 ± 3.8
Abdominal	8.7 ± 5.1	9.7 ± 6.2
Front thigh	16.5 ± 5.6	18.0 ± 5.9
Medial calf	12.3 ± 4.5	13.3 ± 5.2
Mid axilla	5.4 ± 2.0	6.2 ± 3.7[a]
Sum 9 SF[b]	80.3 ± 29.9	90.2 ± 39.7[b]
	Circumferences (cm)	
Head	53.3 ± 1.4	52.6 ± 1.6[a]
Neck	28.0 ± 1.9	26.7 ± 1.4[a]
Arm relaxed	20.1 ± 2.0	19.6 ± 2.3[a]
Arm flexed and tensed	21.7 ± 2.0	21.0 ± 2.4[a]
Forearm	19.8 ± 1.4	19.0 ± 1.6[a]
Wrist	13.6 ± 0.9	12.9 ± 0.8[a]
Chest	68.4 ± 4.7	66.0 ± 6.0[a]
Waist	60.0 ± 4.4	56.4 ± 5.1[a]
Gluteal	71.6 ± 5.5	71.6 ± 6.3
Thigh	42.4 ± 4.2	42.4 ± 4.9
Thigh midtroch-tibiale laterale	39.1 ± 3.5	39.0 ± 4.0
Calf	28.4 ± 2.4	28.3 ± 2.5
Ankle	18.7 ± 1.5	18.3 ± 1.4[a]
	Lengths (cm)	
Acromiale radiale	30.3 ± 1.8	30.0 ± 1.9
Radiale-stylion	22.9 ± 1.5	22.4 ± 1.5[a]
Midstylion-dactylion	16.6 ± 1.1	16.2 ± 1.1[a]
Iliospinale box height	82.7 ± 5.1	81.3 ± 4.9
Trochanterion box height	76.0 ± 4.6	74.9 ± 5.4
Trochanterion-tibiale laterale	38.9 ± 2.7	38.7 ± 2.9
Tibiale-laterale to floor	37.0 ± 2.5	36.5 ± 2.6
Tibiale mediale-sphyrion tibiale	29.3 ± 2.2	29.1 ± 2.1
	Breadths/Lengths (cm)	
Biachromial	31.8 ± 1.9	30.9 ± 2.0[a]
Biiliocristal	21.9 ± 1.6	21.9 ± 1.6
Foot length	22.3 ± 1.6	21.9 ± 1.3[a]
Sitting height	75.7 ± 3.6	74.5 ± 3.8[a]
Transverse chest	21.9 ± 2.6	20.7 ± 1.4
A-P chest depth	15.1 ± 2.2	14.4 ± 2.6[a]
Humerus	6.1 ± 0.4	5.8 ± 0.4[a]
Femur	8.8 ± 0.5	8.4 ± 0.5[a]

[a]Significantly different from boys — $p < 0.05$.

[b]Sum 9 SF — sum of triceps, subscapular, biceps, iliac crest, supraspinale, abdominal, front thigh, medial calf, and mid axilla skinfolds.

Source: Compiled from Jürimäe, T., et al., Influence of anthropometric variables to the whole-body resistance in pre-adolescent children, in Body Composition Assessment in Children and Adolescents, T. Jürimäe and A.P. Hills, Eds., *Medical and Sport Science,* 44, Karger, 2001, 61–70. With permission.

A Tartu-based longitudinal investigation has been conducted to study the possible differences between the anthropometric characteristics of Estonians born and living in Southern Estonia and those of the so-called Russian speakers living in the same region (children whose parents settled in Estonia at least 10 years ago from the former Soviet Union such as Russians, Belorussians, and Ukrainians).[319] The population of Estonia consists of approximately 65% native Estonians and 35% so-called Russian speakers. This investigation followed children from prepubertal stages through puberty. A comparison of anthropometric parameters of boys and girls aged 11 is presented in Tables 2.2 through 2.4.[319]

Table 2.2   A Comparison of Main Anthropometric Variables of 11-Year-Old Native Estonians and Russian-Speaking Children Living in Estonia

Measurement	Boys		Girls	
	Estonians (n = 136)	Non-Estonians (n = 100)	Estonians (n = 125)	Non-Estonians (n = 105)
	Comparative Data			
Stature (cm)	150.3 ± 8.7	148.0 ± 6.4	148.4 ± 6.5	146.8 ± 7.0
Body mass (kg)	40.8 ± 7.2	38.8 ± 7.7	38.4 ± 8.3	37.0 ± 8.5
BMI(kg/m^2)	18.1 ± 1.5	17.7 ± 1.2	17.5 ± 1.1	17.2 ± 1.1
Sitting height (cm)	77.5 ± 3.0	76.7 ± 2.9	77.1 ± 3.5	76.3 ± 3.7

Source: From Jürimäe, T., et al., unpublished data.

The average values of stature, body mass, BMI, and sitting height were slightly higher in native Estonian boys and girls in comparison with Russian-speaking children (Table 2.2). Similar results have also been obtained in another study comparing 10- to 11-year-old Estonian and Russian-speaking children (n = 358) in the northern part of Estonia, in the capital city of Tallinn.[444] These differences among native Estonians and Russian-speaking children may be explained by the cross-cultural difference. Estonians belong to the Fenno-Ugric group, and all Russian-speaking children belong to the Slavic group. Taking genetic potential into account, Estonian children seem

Table 2.3   A Comparison of Circumferential Measurements on the Extremities and the Trunk (cm) of 11-Year-Old Native Estonians and Russian-Speaking Children Living in Estonia

Measurement	Boys		Girls	
	Estonians (n = 136)	Non-Estonians (n = 100)	Estonians (n = 125)	Non-Estonians (n = 105)
	Comparative Data			
Arm	22.5 ± 3.2	22.0 ± 2.7	21.8 ± 3.0	21.6 ± 2.9
Forearm	21.9 ± 1.9	21.5 ± 1.7	20.8 ± 1.7	20.8 ± 1.7
Thigh	45.9 ± 7.2	45.0 ± 5.7	45.9 ± 5.4	46.1 ± 6.7
Calf	30.4 ± 2.8	30.1 ± 2.8	30.2 ± 2.8	30.0 ± 2.9
Chest	72.6 ± 8.5	72.4 ± 6.4	71.8 ± 7.5	70.8 ± 8.1
Waist	64.2 ± 7.8	63.4 ± 6.4	60.9 ± 6.4	60.1 ± 6.4
Gluteal	77.1 ± 9.0	76.3 ± 6.3	78.1 ± 7.3	77.2 ± 7.4

Source: Compiled from Jürimäe, T., et al., unpublished data.

to belong to one of the tallest and heaviest groups in Europe, comparable to the Scandinavian population.[196] Baltic group Lithuanians, in comparison to other nations of the Baltic region, obtained similar values for stature and body mass from prepubertal stages through 17-year-old boys and girls.[313] Most circumferential measurements of native Estonian boys and girls were also higher than in Russian-speaking children (Table 2.3),[319] while length and breadth measurements were almost the same between groups of native Estonians and so-called Russian speakers (Table 2.4).[319]

Table 2.4 A Comparison of Length and Breadth Measurements (cm) of 11-Year-Old Native Estonians and Russian-Speaking Children Living in Estonia

	Boys		Girls	
Measurement	Estonians (n = 136)	Non-Estonians (n = 100)	Estonians (n = 125)	Non-Estonians (n = 105)
		Comparative Data		
Humerus	6.0 ± 0.4	5.9 ± 0.3	5.7 ± 0.3	5.7 ± 0.3
Femur	8.8 ± 0.6	8.8 ± 0.5	8.2 ± 0.5	8.2 ± 0.5
Wrist	5.0 ± 0.4	4.9 ± 0.3	4.7 ± 0.3	4.8 ± 0.3
Biacromial	31.6 ± 1.7	31.1 ± 1.6	30.7 ± 1.9	30.7 ± 1.8
Trochanterion	22.6 ± 1.9	22.2 ± 1.5	22.5 ± 1.7	22.5 ± 1.9
Foot length	23.7 ± 1.4	23.4 ± 1.3	22.9 ± 1.1	22.9 ± 1.3

Source: Compiled from Jürimäe, T., et al., unpublished data.

2.4 Somatotype

The assessment of the physique is most often expressed in the context of somatotype. The somatotype is a description of the morphological state of the individual at a given moment. Numerous attempts have been made to describe the form of the body based on somatotype classification methods. The Heath-Carter method[108,273] is the most commonly used procedure. This method is of particular interest because of the introduction of specific body shape concepts into the definition of three components of the physique. It is expressed in a three-number rating comprised of three consecutive numbers always listed in the same order. Each number represents the evaluation of one of the three basic components of the physique and expresses the individual variations of the human body. In studies of children and adolescent growth, the anthropometric Heath-Carter somatotype method is particularly important because it recognizes that individual somatotypes change over time, most notably during the pre-adult period.[109]

The Heath-Carter anthropometric somatotype method divides the human body into the following components:[108,273]

1. Endomorphy refers to the relative fat of subjects. The endomorphic physique expresses the predominance of digestive organs, softness,

and roundness of contours throughout the body. The endomorphy component characterizes the amount of subcutaneous fat on a continuum from the lowest to the highest values.

2. Mesomorphy refers to the relative musculoskeletal robustness in relation to stature. The mesomorphic physique expresses the predominance of muscle, bone, and connective tissues. Mesomorphy may be considered lean body mass in relation to stature. This component appraises skeletal muscle development on a continuum from the lowest to the highest values.

3. Ectomorphy refers to the relative linearity and fragility of the body. The ectomorphic physique expresses the predominance of body surface area over body mass. Assessment of ectomorphy is based mainly on the index of the ratio of stature to the cubic root of body mass. The lower end of the ectomorphy range describes the relative shortness of various bodily dimensions, while the upper end of the ectomorphy describes the relative length of various bodily dimensions.

Endomorphy, mesomorphy, and ectomorphy are assessed individually. The combined rating of each component describes an individual's somatotype. If one of the components is dominant, the individual's somatotype is described by that component. The extreme values in each component are found at both ends of the scale (a continuum). The low ratings of the endomorphic component signify physiques with a small amount of body fat, while high ratings signify a large amount of body fat. A low value of the mesomorphic component describes an individual with a light skeletal frame and little muscle relief, while a high value of this component implies marked musculoskeletal development. Low values of the ectomorphic component describe subjects with great body mass for a given stature and a low grade of the linearity index (stature/cube root of body mass), while a high value of this component describes a subject with relatively long body segments and little body mass for a given stature together with a high linearity index.[108,273,466,615]

The following anthropometric measures are needed for the anthropometric Heath-Carter somatotype ratings:[108,273]

- Stature, body mass, four skinfolds (triceps, subscapular, suprailiac, calf)
- Two muscle circumferences (flexed arm, calf)
- Two bone diameters (humerus, femur)
- Linearity index (stature/cube root of body mass)
- Age
- Revised linearity index table

The weakness of the anthropometric Heath-Carter somatotyping technique is that it does not use trunk measurements; therefore, differences between limb and trunk segments cannot be evaluated.[108,615] However, this somatotyping

Table 2.5   Estimation of Somatotypes in Estonian Prepubertal Boys and Girls

Measurement	Boys (n = 104)	Girls (n = 105)
	Profile	
Age (yrs)	10.1 ± 0.8	9.8 ± 0.7[a]
Stature (cm)	143.4 ± 7.3	141.5 ± 7.3
Body mass (kg)	35.3 ± 5.7	33.3 ± 6.4[a]
	Skinfolds (mm)	
Triceps	10.0 ± 3.0	11.2 ± 3.9[a]
Subscapular	7.3 ± 3.5	8.4 ± 5.1[a]
Suprailiac	5.1 ± 2.6	6.2 ± 3.8
Calf	12.3 ± 4.5	13.3 ± 5.2
	Circumferences (cm)	
Flexed arm	21.7 ± 2.0	21.0 ± 2.4[a]
Calf	28.4 ± 2.4	28.3 ± 2.5
	Diameters (cm)	
Humerus	6.1 ± 0.4	5.8 ± 0.4[a]
Femur	8.8 ± 0.5	8.4 ± 0.5[a]
Linearity index		
(stature/ $\sqrt[3]{\text{body mass}}$	58.8 ± 6.5	59.0 ± 6.9
Endomorphy	2.2 ± 0.9	2.6 ± 1.2[a]
Mesomorpy	4.2 ± 0.9	3.8 ± 1.0[a]
Ectomorphy	3.5 ± 1.1	3.8 ± 1.3

[a] Significantly different from boys — $p<0.05$.
Source: Compiled from Jürimäe, T. et al., Acta Kinesiologiae Universitatis Tartuensis, 4, 103,

protocol has been widely used because of its lower potential subjectivity, at least in the anthropometric approach.[314,466,615]

The Heath-Carter somatotyping protocol has been used to describe the somatotype components of Estonian 9- to 11-year-old prepubertal boys and girls (Table 2.5).[314] The dominant component in prepubertal children is the mesomorphic one. With regard to sex-linked differences, the endomorphy was significantly higher in girls, while boys presented significantly higher value for the mesomorphy. No significant differences were observed in the ectomorphy component between boys and girls (Table 2.5). The same sex-linked

Table 2.6   Test/Retest Reliability of Heath-Carter Anthropometric Somatotype Components in Estonian Prepubertal Boys and Girls

Variable	Trial I	Trial II	r[a]	SEE[b]
		Boys (n = 24)		
Endomorphy	2.3 ± 1.0	2.3 ± 1.1	0.99	0.2
Mesomorphy	4.1 ± 0.8	4.2 ± 0.9	0.96	0.3
Ectomorphy	3.6 ± 1.1	3.5 ± 1.1	0.97	0.3
		Girls (n = 16)		
Endomorphy	2.6 ± 1.1	2.5 ± 1.0	0.91	0.4
Mesomorphy	3.7 ± 1.0	3.9 ± 0.8	0.89	0.4
Ectomorphy	3.7 ± 1.1	3.8 ± 1.2	0.96	0.3

[a] r — intraclass correlation.
[b] SEE — standard error of estimation.

Source: Compiled from Jürimäe, J., and Jürimäe, T., unpublished data.

differences were also observed in another investigation of 8-year-old (83 boys and 95 girls) and 9-year-old (96 boys and 108 girls) prepubertal children of Estonia.[652] This is in accordance with the results of 8- and 9-year-old children in Hungary[96] and Belgium.[275]

The Heath-Carter anthropometric somatotyping protocol is a reliable method to assess the physique in prepubertal boys and girls. Test/retest reliability values for endomorphy, mesomorphy, and ectomorphy components measured within one week indicated that the intraclass correlation was r > 0.96 with a standard error of estimate (SEE) of <0.3 for boys (n = 24), and r > 0.89 with a SEE of <0.4 for girls (n = 16) (Table 2.6).[311] These results demonstrate that the assessment of the physique using the Heath-Carter anthropometric somatotyping method is an accurate and valid procedure suitable for prepubertal boys and girls.

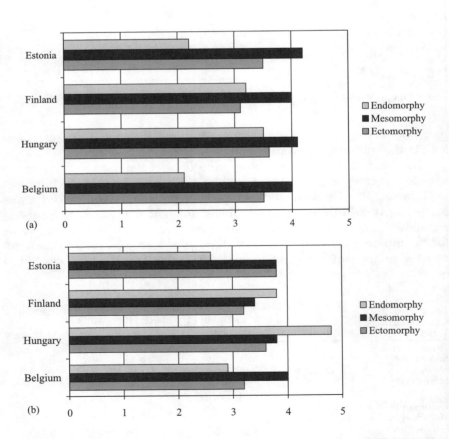

Figure 2.3   Mean somatotypes of prepubertal boys (a) and girls (b) using the Health-Carter anthropometric somatotyping method.

After reviewing growth from infancy to adulthood, Carter and Heath[108] presented a schematic model of the general pathway of children's somatotypes. This model was empirically derived from longitudinal and cross-sectional investigations made in various countries. It is a model that supports and quantifies the concept of somatotype and sex-linked different patterns for both groups and individuals during growth and development.[108,109] Carter and Heath[108] concluded that, in general, males are more mesomorphic and less endomorphic at most ages in comparison with females, while differences in ectomorphy components are less in most studied samples.[108] This was also the case for 9- to 11-year-old prepubertal boys and girls from different countries (Figure 2.3).[96,108,254,275] With regard to sex-linked differences, the average values of the endomorphic components in girls were higher in all children. However, Estonian and Belgian boys and girls had lower values for endomorphy in comparison with Finnish and Hungarian children. All boys had similar and higher values for mesomorphic components than girls. However, the values were markedly lower in Finnish girls in mesomorphic components when compared to girls from other countries. No differences between sexes were observed for ectomorphic components for all children. However, the average values were higher in Estonian and Hungarian boys and girls (Figure 2.3).

According to the model proposed by Carter and Heath,[108] it appears that the physiques of children are fairly stable over some periods of growth. In general, the somatotypes of 2- to 6-year-old children progress from endo-mesomorphy toward balanced mesomorphy for boys and toward central somatotypes for girls up to 6 years of age.[108,109] Thereafter, boys tend to decrease in mesomorphy and increase in ectomorphy up to puberty, when there is a dramatic reversal toward ecto-mesomorphy, balanced mesomorphy, or endo-mesomorphy.[108,109] In prepubertal stages, girls tend toward decreasing mesomorphy, followed by increasing endomorphy, with means moving toward a central somatotype region through puberty and settling somewhere in endo-mesomorphy.[108,109] The overall pattern of change in somatotype components in recent longitudinal studies of Belgian[275] and Canadian[109] children supports the model proposed by Carter and Heath.[108]

2.5   Body composition

2.5.1   The evolution of body composition during childhood

The evaluation of body composition permits the quantification of the structural compartments of the bodies of children. The only direct way to assess body composition in humans is through the dissection and subsequent analysis of cadavers. Consequently, all other assessment methods are indirect. The limited number of cadaver studies, most of which have been on adults, means that data used to formulate reference standards for body composition in prepubertal children are potentially problematic.[283] Many body composition methods are referred to as doubly indirect because they rely on another indirect technique and are subject to the estimation errors inherent in that and subsequent iterations of the data.[283]

Since direct measurement *in vivo* is not possible in humans, a series of indirect estimates of body compartments has been developed. Due to their indirect natures, most of the methods used to measure body composition in humans provide estimations or predictions. Assessment methods range from simple and inexpensive field methods to highly complex and expensive laboratory procedures.[283] To study body composition and changes in body composition during growth, the body mass is subdivided into two or more compartments using elemental, chemical, anatomic, or fluid-metabolic models.[177,278,279,377]

The two-component model, which divides the human body mass into fat and fat-free masses, has generated the most interest. The fat consists of all extractable lipids, while fat-free mass can be divided into water, protein, and mineral compartments.[177,278,377,593] The chemical four-compartment model divides the human body mass into fat, water, protein, and mineral compartments. The human body mass is divided into adipose, non-skeletal muscle soft tissue, skeletal muscle, and bone compartments using an anatomic four-compartment model. The fluid-metabolic body composition model divides the human body mass into fat, extracellular water, intracellular water, extracellular solids, and intracellular solids compartments.[177,278,279,377]

The two-component model is based on the assumption that the fat-free mass consists of 73.8% water, 19.4% protein, and 6.8% mineral, and the density of the fat-free mass is equal to 1.10 $g.ml^{-1}$. The individuals measured differ from each other only in the amount of fat.[279,283,373,377,593] However, prepubertal children are chemically immature. Prior to sexual maturation, children have more water and less bone mineral content than adults.[76,283,287,373,375,377,515,672] The density of fat-free mass steadily increases from 1.063 to 1.102 $g.ml^{-1}$ in males and from 1.064 to 1.096 $g.ml^{-1}$ in females from birth to maturity.[375] In another study, Westrate and Deurenberg[672] estimated that the density of fat-free mass slowly increased with age from 1.080 $g.ml^{-1}$ at 7 years to 1.100 $g.ml^{-1}$ at 18 years in both boys and girls. The water content of the fat-free mass decreases from 79% to 73.8%,[376] and the bone mineral content of the fat-free mass increases from 3.7% to 6.8%[210] in infants to adulthood.

Haschke et al.[265] were the first to define the composition of the fat-free mass of a 9-year-old boy using the data from the literature. The water content of the boy's fat-free mass was calculated to be 75.5%.[263-265] The composition of the fat-free mass of boys during growth and maturation was studied more directly by Haschke,[263,264] when body water was estimated from saliva samples using the deuterium dilution method in 108 boys aged 10 to 15 years. In addition, mineral and protein contents were estimated from prediction equations using body mass and stature.[263,265] It was found that water content of the fat-free mass decreased from 75.2% in 10-year-old boys to 73.6% in 18-year-old boys, indicating that chemical maturity in males was not established until late adolescence.[263] Haschke et al.[263,265] proposed that the 9-year-old reference boy had a mineral content of 4.7% of the fat-free mass. A prepubescent male child has 19.0% of protein content of fat-free mass.[263-265]

Using single-photon absorptiometry, bone mineral content of the distal radius has been measured in 322 children 6 to 14 years old.[417] It was found that the bone mineral content of the distal radius increased about 8.5% per year.[417] In their study, Lohman et al.[380] hypothesized that the bone mineral content of the distal radius and ulna would be associated with body density in children because this regional measure of bone mineral content probably reflected total body bone mineral content. They found significant differences in bone mineral content in four maturation groups from prepubertal stages to adulthood. According to the results of this investigation, Lohman et al.[380] calculated the mineral content of the fat-free mass as 5.3% in prepubertal children, increasing to 6.7% in men and 6.0% in women.

The water content of the fat-free mass is 75.3% in a prepubertal sample of children compared with 72.5% in an adult sample.[76] The decrease in the water content of the fat-free mass during maturation is similar in males (by 2.9%) and females (by 2.8%).[76] Furthermore, a significant gender effect was present, with females having a higher water content than males throughout growth and maturation.[76] The water content of the fat-free mass in adults is 73.5% in males and 74.2% in females.[76,377,378,380]

Table 2.7   Estimations of Age- and Sex-Specific Fat-Free Mass Densities Derived from Multicompartment Models for Children

Gender	7–9 years	9–11 years	11–13 years
Males	1.081	1.084	1.087
Females	1.079	1.082	1.086

Source: Modified from Heyward, V. H. and Stolarczyk, L. M., *Applied Body Composition Assessment,* Human Kinetics, Champaign, 1996; and Lohman T. G., Applicability of body composition techniques and constants for children and youth, in *Exercise and Sport Sciences Reviews,* K. B. Pandolf, Ed., MacMillan, New York, 1986, 325.

The estimations of age- and sex-specific fat-free mass densities derived from multicompartment models for children[279,375] are presented in Table 2.7. The water and mineral contents of prepubertal children are 76.6% and 5.2%,[76,377] respectively, yielding a density of the fat-free mass of 1.084 g.ml^{-1} in prepubertal boys and 1.082 g.ml^{-1} in prepubertal girls. However, according to Brozek et al.,[94] an adult reference man had 73.8% water and 6.8% mineral in the fat-free mass content.

Body composition changes during childhood, and the major changes in the composition of the fat-free mass from prepubertal stages to adulthood occur in the water and mineral compartments. Research has clearly demonstrated that a multicompartment model is necessary for the assessment of body composition from body density in children. Most commonly, multicompartmental models incorporate total body water encompassing extracellular and intracellular water, fat mass, bone mineral, and protein.[283]

Table 2.8   Summary of Body Composition Techniques in Children

Method	Advantages	Limitations	Research needs	Recommendation to use
		Laboratory Methods		
Hydrodensitometry	Directly measures total body density with excellent precision	Not practical for children; difficult maneuver to perform; expensive equipment	Use of air displacement instead of water; determine variability in the density of fat-free mass; more validation research needed	Not recommended for prepubertal children
^{40}K whole body counting	Permits a direct estimation of lean body mass; good accuracy, safe, and non-invasive	Expensive and limited availability; time-consuming procedure		Recommended for prepubertal children
Deuterium oxide dilution	Excellent precision for total body water; could be used as a reference method	Long equilibration time, difficult analysis; expensive		Recommended for prepubertal children
Magnetic resonance imaging and computerized tomography scanning (MRI and CT)	Very accurate; measures tissue area in specific anatomic locations; could be used as a reference method	Expensive and limited availability; CT involves some radiation	Verify representative anatomic sites; develop specific prediction equations to avoid the use of expensive equipment	Recommended for prepubertal children
Total body electrical conductivity	Very accurate; quick and simple; could be used as a reference method	Expensive equipment; limited availability	More validation research needed	Highly recommended for prepubertal children

Table 2.8 (Continued)

Method	Advantages	Limitations	Research needs	Recommendation to use
Dual energy x-ray absorptiometry	Quick and simple; excellent precision for total body bone mineral; capable of regional assessment; could be used as a reference method	Expensive equipment; different machines and software for different subjects	Develop and validate specific prediction equations for children	Highly recommended for prepubertal children
		Field Methods		
Near infrared interactance	Quick and simple; inexpensive; useful for large sample groups	Poor validity and large prediction errors	Development and cross-validation of the technique	Not recommended for prepubertal children
Skinfolds and anthropometry	Quick and simple; inexpensive; useful for large sample groups; accurate for lean participants	Requires high degree of technical skill; very population specific	Development and validation of specific prediction equations for children using the combination of skinfold and anthropometric parameters	Recommended for prepubertal children
Bioelectrical impedance	Quick and simple; inexpensive; useful for large sample groups	Need to have information on hydration status of fat-free mass	Use of multifrequency resistance; development and cross-validation of prediction equations for specific age and ethnic groups	Highly recommended for prepubertal children
Computerized optical system	Precise and simple; inexpensive; useful for large sample groups	Requires high degree of technical skill	More validation research needed	Highly recommended for prepubertal children

2.5.2 Measurement methods

Interest has increased in the assessment of body composition in research during the last decades. Tracking changes in body composition during childhood requires accurate assessments of body compositions of children in laboratory, clinical, and field settings. Many methods (Table 2.8) and prediction equations are currently available for prepubertal children, but none are without limitations. Assessment methods range from simple and inexpensive field methods to highly complex and expensive laboratory procedures. Methods can also be assessed by the degree of skill needed by the tester, the type of equipment required, the degree of cooperation expected of the subject, and the validity and reliability of the method and prediction equation.[283]

The human body can be divided into two chemically-based components, the fat and the fat-free mass. The fat-free mass can be measured from many indirect techniques such as hydrodensitometry, ^{40}K spectrometry, and the deuterium oxide dilution method.[177,283,336] These measurement techniques are based on two assumptions about the composition of the fat and fat-free masses: (1) the fat-free mass composition and density are relatively stable, with little interindividual variability in water, protein, and mineral content; and (2) the composition and density of the fat mass is similar among individuals.[289,374] Research has demonstrated that assumption (1) is not applicable for children (see Chapter 2.5.1). Thus, valid estimates of body composition in children must be based on multi-component models that adjust for deviations from the assumed fat-free mass composition.[248,283,289,374-376,501,515,597] In theory, the more constituents of the fat-free mass that can be estimated, the more accurate the fat content estimates will be. In many research situations, however, only one method is available. In this situation, the use of age- and sex-specific constants is essential for the interpretation of a given body composition method.[210,265,309,374,375,515]

Use of adult values in children substantially overestimates the fat of children when hydrodensitometry,[501,597] predicted density (skinfold thickness method),[309,501] and total body potassium are used,[501] while adult bioelectrical impedance prediction equations tend to uderestimate body fat content in prepubertal boys and girls.[309] Systematic errors of this kind decrease with increasing age.[501,597] However, they are serious and should be avoided. The use of the age- and sex-specific constants approach removes systematic errors from pediatric body composition methods and produces robust methods that are acceptably accurate.[287,309,377,501,597]

One of the major limitations of comparing body composition techniques in children is the lack of a reference standard method. Hydrodensitometry, ^{40}K whole body counting, and deuterium oxide dilution methods are commonly accepted and utilized validation criteria for checking field measures.[177,283,336] The use of traditional hydrodensitometry as a criterion method in children is limited because many children have difficulties with the breathing maneuver involved in the determination of underwater weight.[336,377] The procedure of total submergence also presents difficulties for children who are inexperienced or highly anxious in water. Total body potassium assessment requires total body enclosure in a cylindrical chamber for several minutes while remaining completely still. This is also a difficult task for most young children.[152,336] The deuterium oxide dilution method appears to be the most suitable criterion for prepubertal children because it only requires that the child ingest a 5-gram dose of deuterium oxide and provide urine samples for spectroscopic assay.[152,336,514] In addition, dual-energy x-ray absorptiometry has been used as a reference standard body composition technique in prepubertal children.[154,283]

To determine the criterion variables of body composition in children, multiple component body composition models have been suggested.[283,288,374,375,501,515] There is a relatively large variability in the proportions of the various components of the fat-free mass in prepubertal children, particularly the water and bone mineral contents. To obtain accurate measures, body water and bone mineral contents and body density should be measured separately to determine the criterion variables of body composition and assess the reliability and accuracy of field methods. For example, a commonly used multiple component body composition model approach adjusts the body density obtained from hydrodensitometry for total body water from deuterium oxide dilution and bone mineral from dual energy x-ray absorptiometry.[152,660]

Hydrodensitometry (underwater weighing) estimates body composition from measurement of total body density. The most widely used approach is to measure body volume by underwater weight and determine body density by dividing body mass by body volume. Densitometry with simultaneous measurements of air in the lungs and respiratory passages is a simple and reliable method.[204,279,283,377,660] It has been used for the calibration of nearly all new methods. However, it is not recommended to use this method in prepubertal children because it demands a great degree of cooperation and lack of fear of submerging the head underwater. The subjects must be able to exhale completely and hold their breath for at least 10 seconds underwater, and this must be repeated several times. Although Parizkova[466] succeeded in measuring body density using underwater weighing in children aged 6 to 7 years old, we were not able to obtain reliable data in 9- to 11-year-old boys and girls with underwater weighing.[311]

Recent developments using air rather than water displacement for the measurement of body volume may be more practical for pediatric populations.[237] This device, called BodPod®, is simpler, faster and more practical than underwater weighing.[237,283,660] This method is based on air displacement plethysmography and uses the relationship between pressure and volume to derive body volume for a subject seated inside the chamber. Body volume is equal to the volume of air in an empty chamber minus the volume of air remaining in the chamber after the subject has been placed into it.[660] Recently Field and Goran[205] compared the accuracy, precision, and bias of fat mass as assessed by dual-energy x-ray absorptiometry, hydrostatic weighing, air displacement plethysmography using the BodPod, and total body water against the four-compartment model in prepubertal children (11.4 ± 1.4 years). They concluded that air displacement plethysmography was the only technique that could accurately, precisely and without bias estimate fat mass in 9- to 14-year-old children.[205]

^{40}K whole body counting estimates total body potassium, which serves as an indicator of fat-free mass. There is very little potassium outside cells, and intracellular concentration of potassium is relatively constant and distributed entirely within the fat-free mass.[177,179,283] Total body potassium can be measured in two ways — the subject can be placed in a chamber that counts the decay of ^{40}K, a naturally radioactive isotope, or the ^{40}K can be administered to the subject. This option does not require the extensive shielding and sensitive detectors required in a chamber. Based on the radioactivity measured from the counting procedure, ^{40}K can be measured and converted into lean body mass using a factor dependent on the potassium content of the fat-free mass.[283] ^{40}K spectrometry has been used simultaneously with skinfold thickness and bioelectrical impedance analysis measurements in subjects 4 to 19 years old.[179] This technique is safe and non-invasive; however, the need to be enclosed in a small chamber for a period of time may limit the use of this technique in young children.[283]

The deuterium oxide dilution method is a comparatively safe and valid approach to assess body composition in children. Deuterium is a stable isotope and has been used extensively in children.[145,466] Orally ingested, deuterium oxide is readily absorbed in the gastrointestinal tract and is in equilibrium with body water within a few hours. The equilibrium concentration can be determined in blood, urine, or saliva. Infrared absorption, falling drop method, freezing point elevation, mass spectrometry, and gas chromatography have been used to measure deuterium oxide concentration.[466] This technique has been widely used for body composition measurements in prepubertal children.[145,146,152,466] The deuterium oxide dilution method has also been used as a reference method in a sample of prepubertal children.[145,152,336]

Recently, *in vivo* imaging techniques (magnetic resonance imaging [MRI] and computerized tomography [CT]) have allowed more accurate measures of body composition in children. These imaging techniques give a two-dimensional image of the human body at the scanned body level. Analyzing the scans allows the calculation of the cross-sectional area of tissues at the measured level. Making more scans of the human body allows the calculation of the volume of the tissues;[594] and when the density of tissue is known, the amount in kilograms can be calculated.[160] The advantage of MRI and CT is that these techniques are able to depict internal and visceral fat and subcutaneous fat along with other tissues. The assessment of body fat distribution is equally as important as measurement of total body fat because intra-abdominal adipose tissue is related to negative health outcomes independent of total body fat.[62,237] The main limitation of MRI and CT is that the subject must be enclosed in the scanner for a long period of time — which may be traumatic, especially for younger children. These methods are expensive, time consuming, and not accessible for all practitioners. They therefore have limited application for work with pediatric populations.[283]

Total body electrical conductivity (TOBEC) is another highly suitable method for body composition assessment in children.[74,283,466,501] This technique is based on the differences in electrical properties of fat and fat-free mass. The measurement chamber of the TOBEC apparatus consists of a large cylindrical coil. Alternating current in the coil creates an electromagnetic field that induces an opposing current within the human body. Energy is absorbed by the human body and released as heat. The absorbed energy is measured by a decrease in coil impedance. The change in impedance is related to the dielectric and conductive properties of the human body; and prediction equations for water, fat, and fat-free mass are developed. The TOBEC technique is considered to be a highly reliable and relatively accurate means of estimating total body water and fat-free mass.[74,283,466,501] The TOBEC method has been used to measure fat-free mass in prepubertal children.[74,283,649]

Dual energy x-ray absorptiometry (DEXA) is a relatively new technology that is gaining recognition as a reference method for body composition research. This method uses x-rays of two different energy levels,[283,418] yielding a one-dimensional scan of the human body from which the total and the segmental (arm, leg, trunk) estimates of bone mineral, fat-free tissues, and fat tissue can be obtained. DEXA requires only 10 to 15 minutes for a whole-body analysis. The minimal radiation exposure allows repeated studies. Since the introduction of DEXA, numerous studies have compared this technique with other research-based methods.[250,485,650] These investigations have proven valuable in cross-validating DEXA. A general limitation of many widely used body composition techniques is the lack of validation studies that perform comparisons with known chemical compositions.[237]

The validity of the DEXA model has been demonstrated by comparison with chemical analysis of the carcass of a pig.[180,237,466,484] The results of DEXA were significantly related to the total carcass analysis in pigs with a weight range of 5 to 35 kg.[180,237,484] However, significant differences in the partitioning among bone mineral content, non-bone mineral tissue, and body fat compartments were revealed.[180] The DEXA method using adult scan analysis failed to measure body composition accurately in a group of boys aged 4 to 12 years.[180] Another investigation in the pediatric body mass range using pigs demonstrated that the lean and fat contents of the carcass were highly correlated with the DEXA measurements ($r > 0.98$).[484] The regression between the carcass and DEXA data deviated by a small but significant amount.[484] This suggests that specific correction factors may need to be used to improve the measurement accuracy of total body composition by the DEXA pig model.[180,237,484] The validation and development of new calibration equations are important steps in the development of standardized techniques to measure body composition in children.[237,283,484,650]

The near-infrared interactance (NIR) method is a relatively new field method to estimate body composition. The NIR technique is based on the principle that the pattern of near infrared light that is reflected by fat and fat-free tissues varies. NIR relies on the principle of light absorption and reflection. An infrared light beam is placed over the biceps muscle, and reflected energy from a fiberoptic probe is monitored by an optical detector. The NIR method compares the light absorption properties of two wavelengths. In combination with other anthropometric data, NIR predicts body composition utilizing an appropriate regression equation. Very few prediction equations have been presented for children, and these equations are characterized by unacceptable prediction errors.[113,278,279,283] At the present time, there is limited research to estimate body composition in children using the NIR method. Although the Futrex-5000™ NIR analyzer (Model 5000 A) includes prediction equations for children 5 to 12 years old and adolescents 13 to 18 years old, the cross-validation of these equations indicates a systematic overestimation of the average percent of body fat by 2.5 to 4.1%.[113] Therefore, more work is required to cross-validate the technique before using it for the assessment of body composition in prepubertal children. The current prediction equations for children are not recommended for body composition assessment. This technique may have application in the future as an appropriate field method since NIR requires little technical skill, little subject cooperation, and minimal equipment.[283]

Skinfolds and anthropometry involve development of prediction models in which anthropometric measures are related to body fat mass. The skinfold method indirectly measures the thickness of subcutaneous adipose tissue. The value for subcutaneous fat assessed by skinfold measurements at 12 sites is similar to the value obtained by MRI,[271] and skinfold thicknesses at multiple sites of the human body measure a common human body fat factor.[271,278,283,298] Furthermore, it is assumed that approximately one third of total body fat is located subcutaneously in men and women.[278,373] Considerable biological variation exists in subcutaneous, intramuscular, intermuscular, and internal organ fat deposits as well as essential lipids in bone marrow and the central nervous system.[278,373] This biological variation in body fat distribution is affected by age, gender, and amount of fat.[373]

Skinfold measurements are relatively simple, and their validity and reproducibility are high when taken properly. Skinfold prediction equations are based on linear (population-specific) or quadratic (generalized) regression models. There are over 100 population-specific equations to predict body density or percent of body fat from various combinations of skinfolds, circumferences, and bone diameters.[204,283,299] These equations have been developed for homogeneous populations. Generalized equations are used in individuals varying greatly in age and body fat.[211,279,283,373] These equations also take into account the effect of age on the distribution of subcutaneous fat. The use of an accurate criterion method is important for the development of these equations.

Several prediction equations have been developed and cross-validated for children based on the use of a multicompartmental model,[252] deuterium oxide dilution,[26,145,152,336] or DEXA[154,240] as a criterion method.

Slaughter et al.[597] developed body composition equations from data on 310 subjects (8 to 29 years of age), including 66 prepubertal children (50 boys and 16 girls). A multicompartmental model of criterion method included three separate approaches to measure total body density (underwater weighing), total body water (deuterium oxide dilution) and bone mineral density (photon absorptiometry) on the right and left radius and ulna. In the prepubertal group of children, systematic differences were found among methods, with the underwater weighing alone producing higher mean body density than the other two estimates. This fact again supports the concept that constants used to estimate fat mass in adults tend to overestimate body fat in children.[597] The Slaughter et al.[597] research led to the development of gender-, race-, and maturation-specific prediction equations for estimating body density based on the measurement of either triceps and calf or triceps and subscapular skinfolds. In the prepubertal group, the triceps and calf skinfold combination yielded a coefficient of determination of 77% with a SEE of 3.9%; for the triceps and subscapular combination, the coefficient of determination was 80% with a SEE of 3.6%.[597]

Several studies have examined the accuracy of the Slaughter prediction equations for prepubertal children.[154,240,303,309] Janz et al.[303] cross-validated the Slaughter equations based on the sum of triceps and calf skinfolds in prepubertal boys and the sum of triceps and subscapular, or sum of triceps and calf skinfolds in prepubertal girls. Underwater weighing was used to determine total body density, and the percent of body fat was obtained using Lohman's[375] age-gender conversion formulas.[303] For girls, both equations had acceptable prediction errors (SEE = 3.5 to 3.6% body fat). However, the sum of triceps and calf skinfold prediction equations slightly overestimated (by 1.7%) the average percent of body fat in girls. For boys, the prediction error for the sum of triceps and calf skinfold equations was unacceptable (SEE = 4.6% body fat) and varied with maturation level.[303] In a cross-validation investigation by Goran et al.,[240] the fat mass measured by DEXA (4.8 ± 3.0 kg) was significantly lower than fat mass estimated by subscapular and triceps skinfolds with the Slaughter prediction equations (5.0 ± 3.1 kg), although fat masses by these two body composition measurement methods were strongly related ($R^2 = 0.87$; SEE = 1.1 kg).[240] The differences between equations predicting fat mass may be explained by the small sample of prepubertal girls (n = 16) in the Slaughter et al.[597] study as well as the slightly younger age of the subjects (7 vs. 10 years) in the Goran et al.[240] investigation.

Measures such as BMI and sum of skinfolds are recommended as indices of body composition for prepubertal children without expressing the values as percent of body fat.[26,152,250,283,309] For example, Gutin et al.[250] found a high correlation between the sum of seven skinfold measurements and the percent of body fat derived from the two skinfold equations of Slaughter et al.[597] in prepubertal children (r = 0.97). In another investigation, Bandini et al.[26] found that triceps skinfold alone and BMI characterized body fat percent measured using deuterium oxide dilution method by 68% and 38%, respectively, in pre-menarcheal girls aged 8 to 12 years (n = 132).

In our recent study, the sum of triceps and calf; the sum of chest, abdominal and mid-thigh; the sum of triceps, biceps, subscapular, suprailiac and calf; and the sum of triceps, subscapular, chest, midaxillary, suprailiac, mid-thigh and calf skinfolds were closely related to the percent of body fat estimated from different skinfolds and bioelectrical impedance analysis regression equations in boys and girls 9 to 11 years old.[309] The computation of the sum of triceps and calf skinfolds is a simple way to monitor changes in body composition of prepubertal boys and girls.

Our study compared different skinfold thickness prediction equations found in the literature in 212 prepubertal boys and girls (Table 2.9).[309] Body density was calculated using Jackson and Pollock[299] and Jackson et al.[300] generalized equations developed on adult samples. Percent of body fat was then calculated using Lohman's[373] age-specific constants. For comparison, percent

Table 2.9 Mean (±SD) and Ranges of Percentage Body Fat (%BF) Estimated from Skinfolds (SF) and Bioelectrical Impedance Analysis (BIA) Regression Equations

Dependent variable	Boys (n = 107)	Girls (n = 105)
SF (Lohman)[a]	11.0 ± 3.9[g]	14.7 ± 4.3[j]
SF (Siri)[b]	16.7 ± 4.0	19.7 ± 4.7[k]
SF (Slaughter)[c]	15.0 ± 4.9[h]	19.6 ± 5.0[k]
SF (Boileau)[d]	12.5 ± 4.6[g,i]	17.3 ± 5.2[l]
BIA (Deurenberg)[e]	19.7 ± 4.9	25.1 ± 5.2
BIA (Houtkooper)[f]	13.7 ± 4.8[h,i]	16.4 ± 5.5[j,l]

[a]SF technique using Jackson and Pollock[211] and Jackson et al.[276] regression equations for boys and girls, respectively, to calculate body density (BD) and Lohman's[206] age-specific constants to calculate %BF.

[b]SF technique using Jackson and Pollock[211] and Jackson et al.[276] regression equations for boys and girls, respectively, to calculate BD and the Siri[204] equation to calculate %BF.

[c]SF technique using the Slaughter et al.[597] age-specific equation to calculate %BF.

[d]SF technique using the Boileau et al.[75] age-specific equation to calculate %BF.

[e]BIA technique using the Deurenberg et al.[256] age-specific equation to calculate fat-free mass and subsequently %BF.

[f]BIA technique using the Houtkooper et al.[235] age-specific equation to calculate fat-free mass and subsequently %BF.

[g,h,i,j,k,l] SD in the same column with the same superscript are not significantly different (p > 0.05). All other ±SD are significantly different (p < 0.05).

Source: Modified from Jürimäe, J., et al., *Med. Dello Sport*, 51, 341, 1998. With permission.

of body fat was calculated from body density using the adult equation developed by Siri.[593] Percent of body fat was also derived using Slaughter et al.[597] and Boileau et al.[75] age-specific prediction equations. The lower values in both groups were obtained using Jackson et al.,[300] Jackson and Pollock[299] body density equations, and Lohman's[373] age-specific constants. Percent of body fat in boys calculated from the age-adjusted equations of Slaughter et al.[597] and Boileau et al.[75] was significantly lower than the percent of body fat values obtained using the adult equation of Siri.[593] In contrast, the percent of body fat values obtained using the Siri[593] equation and Slaughter et al.[597] age-specific equation were similar. These results are in accordance with Janz et al.,[303] who reported that the triceps and calf skinfold equations of Slaughter et al.[597] tended to systematically overpredict percent of body fat in girls. Thus, the accuracy of percent of body fat estimation in children depends on the selection of an appropriate prediction equation.[154,303,309]

The bioelectrical impedance analysis method is an appealing tool for *in vivo* assessment of body composition because it is simple, fast, and inexpensive to perform. Theoretically, the bioelectrical impedance method is based upon the relationships among the volume of the conductor (the human body), the conductor's length, the components of the conductor, and the conductor's impedance. It is assumed that the total conductive volume of the human body is equivalent to that of total body water, most of which is contained in muscle tissue, and that the hydration of adipose tissue is minimal.[42,160,279,283,284,515] Total body impedance, measured at the constant frequency of 50 kHz, primarily reflects the volumes of water and muscle compartments comprising the fat-free mass and the extracellular water volume.[42,117,204,279,284,515] The resistance to current flow is greater in individuals with large amounts of body fat since adipose tissue is a poor conductor of electrical current due to its relatively small water content.[160,279,284,515] Because the water content of fat-free body mass is relatively large (73% water), fat-free mass can be predicted from total body water estimates. Individuals with a large fat-free mass and total body water have less resistance to current flowing through their bodies in comparison with persons having a smaller fat-free mass.[160,279,284,515]

However, the intracellular penetration is not complete at the frequency of 50 kHz. Since the cell membrane behaves as an electric capacitor, alternating currents at low frequency are not able to penetrate the cell. Thus, at low frequency, the impedance of the human body is a measure of intracellular water. With increasing frequency, the reactance of the cell membrane decreases and finally disappears. Accordingly, at high frequency, bioelectrical impedance is a measure of total body water.[117,158,160,161] Measures of bioelectric impedance at higher frequencies have been reported to discriminate the volumes of intracellular and total body water in the human body.[42,117,158,160,161,279,284,515] Differences in the distribution of fluids between intra- and extracellular compartments, which occur during growth and development, could help to explain the variability in the prediction of fluid status or change in fluid status in children. It is important to measure the extracellular water compartment with bioelectrical impedance because extracellular water volume is about 20 to 30% of body mass, and changes in extracellular water volume also occur in malnutrition.[117]

The new bioelectrical impedance instruments can measure body impedance at more than one frequency, ranging from low (about 1 kHz) to very high (> 1 mHz).[117,156,160] At low frequency, body impedance is a measure of extracellular water; and at high frequency, body impedance is a measure of intracellular water. Multifrequency impedance analyzers can be used to monitor changes in fluid status in children since the variation in hydration of fat-free mass is relatively high in children.[210,277] For example, the percentage of body water in boys from birth to 10 years of age decreases as does the ratio of extra- and intracellular water.[210]

Total body water and fat-free mass are significantly related to the stature squared divided by resistance (S^2/R).[42,135,145,152,248,279,284,290] As in adults, the measurement of body composition in prepubertal children using the bioelectrical impedance analysis method utilizes this resistance index (S^2/R) in different regression equations. Most researchers confirm that the presented index is applicable for the calculation of different body composition parameters.[42,135,145,284,355,377,387] However, some researchers[152,248,288,316,460] have recommended that additional anthropometric parameters to stature be used in prediction equations. In addition to the resistance index, the independent variables used most often by investigators in their prediction equations are body mass, arm circumference, sex, and age.[117] The main problem is that the stature is not the correct length of the conductor. The true length of the conductor is better represented by the acromial stature and arm length.[116,247,314] In our recent study,[316] stature alone characterized only 1.9% (p > 0.05) and 3.8% (p < 0.05) in body resistance of the total variance in prepubertal boys and girls, respectively. The better predictor of body resistance is body stature and mass combined (27.1% and 20.7%, respectively).[316]

Rather than stature alone, additional anthropometric measures are needed to present new prediction equations for calculation of body composition in children. A study of young adults, using body mass, upper arm and calf circumferences, and seven skinfold thicknesses, revealed that about 70% of the variance in body resistance could be accounted for by a small set of anthropometric variables such as arm and calf circumferences.[38,42] Significant correlations with body resistance exist for body mass, upper arm and calf circumferences, upper arm and calf muscle areas, ratios of limb segments to their lengths, and some skinfold thicknesses.[38,42,284,287,355] Our recent investigation of boys and girls 9 to 11 years old also indicated that the best predictors of body resistance were girth parameters, which characterized about 30 to 50% of the total variance in body resistance.[316] The variance occurs because the cross-sectional area of the human body is not constant, and the parts with the smallest cross-sectional areas primarily determine the resistance of the human body.[42] However, it is interesting to note that small limb girths and gluteal and waist girths in boys and girls, respectively, were also added to the prediction model. Girth ratios such as waist/hip and waist/thigh have been used by most investigators as a measure of fat distribution with variable results. In our recent study in prepubertal children,[316] correlations between body resistance and waist/thigh ratio were not only moderate but significant in boys and girls. Both indices are significantly related to the impedance index in adults.[460] The waist/thigh ratio is more important because this ratio contains the girth of lower limbs in which body resistance is relatively high.

As in adults,[42,355] body length parameters only slightly influence body resistance in prepubertal children.[316] This is surprising because the body resistance depends on the conductor length. The very small girth of the upper and lower bodies in children is a potentially higher predictor than the length of the limbs. The influence of skinfold thicknesses to body resistance was also found to be low in prepubertal children, characterizing less than 10% of the total variance.[316] The sum of skinfolds characterized 7.2% of the body resistance in girls and 2.4% in boys[316] because body fat is a very poor electric conductor.[42,283,335,386] Traditional body stature as a single anthropometric measure used in the presentation of equations for body composition measurement in children is not acceptable. It is important to add girth parameters to body stature in the prediction of body composition in prepubertal children.

The influence of somatotype on body resistance in prepubertal children has also been studied.[314,315] The impact of ectomorphy on body resistance was significant in boys ($r = 0.33$ to $r = 0.48$) and girls ($r = 0.21$ to $r = 0.43$), while the impact of endomorphy on body resistance was not significant ($r < -0.18$) in boys and partly significant ($r = -0.19$ to $r = -0.30$) in girls. The mesomorphic component negatively influenced the body resistance in boys ($r = -0.49$ to $r = -0.65$) and girls ($r = -0.31$ to $r = -0.45$).[314,315] The high correlation between body resistance and the mesomorphic component is not surprising since this somatotype component characterizes the relative musculoskeletal robustness of the human body and is derived from biepicondylar femur and humerus widths as well as arm and calf circumferences corrected for skinfolds.[108] The thinner segments of the body provide the greatest resistance when they are also long.[223] Regression analysis predicting body resistance indicated that only the mesomorphic component in boys (45.8%) and the mesomorphic and ectomorphic components combined in girls (51.3%) were significant predictors of body resistance.[314,315] According to the results, it is apparent that the relative robustness, and the relative linearity and robustness, are the components that greatly influence body resistance in 9- to 11-year-old boys and girls, respectively.

Table 2.10   Resistances (Ω) and Volumes (l) in Total Body Water (TBW), Intracellular Water (ICW), and Extracellular Water (ECW) Measured at Different Sites of the Body in Boys and Girls

Site/Method	Boys (n = 104)	Girls[a] (n = 105)
Right Side		
5 KHz (Ω)	622.4 ± 65.0	671.1 ± 68.9
50 KHz (Ω)	578.8 ± 58.3	626.8 ± 56.6
200 KHz (Ω)	522.8 ± 53.6	564.2 ± 50.6
TBW (l)	24.3 ± 2.6	20.1 ± 2.9
ICW (l)	12.3 ± 1.4	8.9 ± 1.2
ECW (l)	12.0 ± 1.3	11.3 ± 1.2
Left Side		
5 KHz (Ω)	637.5 ± 66.0	692.9 ± 69.4
50 KHz (Ω)	595.4 ± 61.3	644.0 ± 60.6
200 KHz (Ω)	540.8 ± 58.5	587.3 ± 54.0
TBW (l)	23.8 ± 2.5	20.0 ± 2.3
ICW (l)	12.0 ± 1.5	8.8 ± 1.3
ECW (l)	11.9 ± 1.2	11.1 ± 1.1
Hand-Hand		
5 KHz (Ω)	687.8 ± 74.4	759.9 ± 83.6
50 KHz (Ω)	650.3 ± 71.0	713.5 ± 72.5
200 KHz (Ω)	592.6 ± 65.2	653.3 ± 64.6
TBW (l)	23.1 ± 2.6	19.1 ± 2.2
ICW (l)	11.5 ± 1.7	8.5 ± 1.1
ECW (l)	11.6 ± 1.3	10.7 ± 1.3
Leg-Leg		
5 KHz (Ω)	525.9 ± 53.5	581.3 ± 61.9
50 KHz (Ω)	485.1 ± 48.9	532.5 ± 60.0
200 KHz (Ω)	439.1 ± 47.3	480.5 ± 58.2
TBW (l)	25.9 ± 3.4	22.0 ± 2.7
ICW (l)	12.9 ± 2.1	9.8 ± 1.4
ECW (l)	13.0 ± 1.5	12.2 ± 1.4
Right Hand–Left Leg		
5 KHz (Ω)	635.8 ± 67.3	711.1 ± 72.8
50 KHz (Ω)	592.7 ± 60.4	657.5 ± 65.0
200 KHz (Ω)	539.2 ± 55.8	595.9 ± 59.9
TBW (l)	23.8 ± 2.7	19.9 ± 2.4
ICW (l)	11.9 ± 1.8	8.8 ± 1.3
ECW (l)	11.9 ± 1.2	11.1 ± 1.5
Left Hand–Right Leg		
5 KHz (Ω)	632.4 ± 68.1	702.8 ± 69.5
50 KHz (Ω)	592.4 ± 61.7	650.8 ± 62.6
200 KHz (Ω)	539.2 ± 57.1	593.3 ± 51.6
TBW (l)	24.0 ± 2.7	19.9 ± 2.3
ICW (l)	12.0 ± 1.7	8.8 ± 1.2
ECW (l)	12.0 ± 1.3	11.0 ± 1.2

[a]All values are significantly different from boys, $p < 0.001$.

Source: Compiled from Jürimäe, J., et al., Whole-body resistance measured between different limbs and resistance indices in pre-adolescent children, in Body Composition Assessment in Children and Adolescents, Jürimäe, T., and Hills, A.P., Eds., *Medical Sport Science,* 44, 2001, 53–60.

Our recent study compared the results of body resistance measured at different sites of the body in boys and girls 9 to 11 years old.[310] We hypothesized that the best approach was to measure between the right leg and the left hand, or between the left leg and the right hand, for true measurement of body resistance. In addition, the possible differences in body resistance were compared when body resistance was measured traditionally between arm-to-leg on the right or left side of the body or between lower and upper extremities in prepubertal children.[310]

In all cases, the mean body resistance was significantly higher in girls than in boys (Table 2.10). The mean difference between right and left side measurements at 50 kHz was 16.6 Ω (2.8%) and 17.2 Ω (2.7%) in boys and girls, respectively.[310] The resistance is systematically greater on the left side than on the right side of the body.[244,288,290,310] For example, Graves et al.[244] found that resistance was about 8 Ω greater on the left side of the adult human body. The side on which resistance is measured must be the side where the body resistance was measured during the development of the body composition predictive equation. The values of body resistance measured diagonally (right hand–left leg or left hand–right leg) were comparable with right-side body resistance measurements in both groups (Table 2.10).[310] The results of this investigation did not confirm the hypothesis that it is more precise to measure body resistance diagonally between hand and opposite leg than on the right side of the body. The measurement of body resistance on the right side of the body is correct in prepubertal children.[310]

The lower and upper body resistances were higher than the whole body resistance (right arm–trunk–right leg) because of the relatively smaller volumes of these body segments in comparison with the trunk.[310] The measured body resistance between hands was significantly higher than when measured between legs (Table 2.10),[310] likely because the breadths of the hands are slightly shorter than those of the legs. The thinner segments of the body provide greatest resistance.[38,223,244,278,460]

Segmental body impedance, which refers to the measurement of body resistance of different body segments, is important in the estimation of regional body composition.[41,42,136,283,460,515] The resistance is larger for the parts of the human body with the smallest circumferences.[41,42,177,283,460] For example, the arm contributes to only about 4% of body mass but as much as 45% to the resistance of the whole body.[165,177,223,460] In contrast, the trunk, which has a large cross-sectional area, contributes to about 46% of body mass but is responsible for only about 11% of the whole body resistance.[165,177,223,460] Electrode placements for the separate measurement of the major body segments (arm, leg, trunk) have been described,[116,136,223,460] but a standardized procedure has not yet been recognized.[136,155] The measures of bioelectrical impedance are larger in women than in men for the limbs, while the sex differences were not significant for the trunk region of the body.[515] The same pattern of sex differences occurs in children, except that the bioelectrical impedance of the trunk is larger in boys than in girls.[40]

The positioning of the electrodes is important for both whole body and segmental bioelectrical impedance measurements.[42,136,290] The displacement of the source electrodes proximally by 1 cm, on either the hand or the foot, reduces the measured resistance by 2.1%.[42,181,561] Interobserver differences associated with the placement of electrodes can be reduced when the sites of electrode placement are marked.[42,181] However, when the source and receiving electrodes are placed closer together than 4 to 5 cm, electron polarization may occur that will increase the resistance.[42] This problem may limit the use of bioelectrical impedance methods to children.[42] For example, in their study with children 3 to 10 years old, Barillas-Mury et al.[32] were able to separate the electrodes sufficiently to stabilize the resistance on the feet but not on the hands. To solve this problem with the hand, the researchers placed one signal electrode on the dorsal wrist and one source electrode on the dorsal aspect of the forearm 6 cm proximal to the wrist.[32] Placement of electrodes is critical to obtain accurate bioelectrical impedance measurements in children.

Age-specific prediction equations have also been recommended for bioelectrical impedance analysis.[42,152,157,159,237,283,289,502,515,560] Age-related differences in the electrolyte concentration in extracellular water space relative to intracellular water space may alter the relationship between bioelectrical resistance and total body water.[159] However, Houtkooper et al.[289] reported that including age as a predictor did not significantly improve the predictive accuracy of the bioelectrical impedance analysis equation. The prediction formula of Houtkooper et al.[289] for white boys and girls was developed using a three-component model that adjusted body density for total body water. This prediction equation has been cross-validated on samples from three different laboratories with a prediction error of 2.1 kg.[289,502]

Other bioelectrical impedance prediction equations recommended for use with prepubertal children have been developed by Guo et al.[248] These regression equations to predict the fat-free mass for males and females had R^2 values of 0.98 and 0.95 and SEEs of 2.3 and 2.2 kg, respectively.[248] The retained predictor variables were body mass, calf and midaxillary skinfolds, S^2/R index, and arm circumference in males.[248] For females, the retained predictor variables were body mass, calf, triceps and subscapular skinfolds, and S^2/R index.[248] These equations did not overpredict or underpredict for different parts of the distribution of values for fat-free mass.[248] These regression equations have been used in Fels' longitudinal study to predict percent of body fat in prepubertal children.[515] In their study, Evetovich et al.[197] evaluated the validity of 11 existing bioelectrical impedance equations in a group of 11-year-old male sportsmen (n = 117) and found that the equation of Guo et al.[248] most accurately estimated the fat-free mass of subjects (SEE = 1.99 kg).

The bioelectrical impedance technique has been cross-validated in children against DEXA,[240,303,458] total body water,[42,238] and total body potassium[42,560] methods. Schaefer et al.[560] estimated the fat-free mass from ^{40}K whole body counting in 112 healthy children and demonstrated that the fat-free mass could be estimated from bioelectrical impedance and age with an R^2 value of 0.98. In another independent cross-validation investigation in 98 white children, the fat mass measured by DEXA (4.8 ± 3.0 kg) was significantly different from the fat mass measured by bioelectrical resistance (5.7 ± 3.4 kg), although the fat masses by these two techniques were strongly related ($R^2 = 0.75$).[240]

Our recent investigation compared the results of percent of body fat using two regression equations developed for children by Deurenberg et al.[159] and Houtkooper et al.[289] in 107 prepubertal boys and 105 prepubertal girls (Table 2.9).[309] The Deurenberg et al.[159] prediction equation yielded considerably higher estimation of body fat in comparison with the Houtkooper et al.[289] equation for both groups.[309] Furthermore, the mean percent of body fat from the latter prediction equation did not differ from those obtained using the skinfold prediction equations of Slaughter et al.,[597] Boileau et al.,[75] and Lohman[373] for boys and girls. The considerably higher values for percent of body fat using the Deurenberg et al.[159] prediction equation could be explained by the fact that they used only pubertal children and found relatively poor reproducibility. The choice of a bioelectrical impedance analysis regression equation to estimate body composition in prepubertal children is critical, even from bioelectrical impedance prediction equations developed specifically for children.

Table 2.11 Test-Retest Reliability of Total (TBW), Extracellular (ECW) and Intracellular (ICW) Body Water Measurements in Estonian Prepubertal Boys and Girls

Variable	Trial I	Trial II	r^a	SEE[b]
		Boys (n = 24)		
TBW (l)	24.8 ± 2.4	25.0 ± 2.5	0.96	0.7
ECW (l)	12.5 ± 1.2	12.3 ± 1.5	0.88	0.7
ICW (l)	12.7 ± 1.3	12.7 ± 1.1	0.92	0.4
		Girls (n=16)		
TBW (l)	20.1 ± 2.2	20.3 ± 2.7	0.97	0.7
ECW (l)	11.4 ± 1.2	11.2 ± 1.4	0.93	0.5
ICW (l)	8.7 ± 1.0	8.9 ± 1.2	0.99	0.2

[a] r — intraclass correlation.
[b] SEE — standard error of estimation.
Source: Compiled from Jürimäe, J., and Jürimäe, T., unpublished data.

Substantial evidence suggests that bioelectrical impedance measurements are highly reliable based on interobserver and intraobserver comparisons with replacements of electrodes as well as on interday and interweek comparisons.[42,250,283,301,355,387,687] Test/retest reliability data for total, extracellular, and intracellular body water values indicated that, for prepubertal boys (n = 24) and girls (n = 16) measured within one week, the intraclass correlation was r > 0.88 with a SEE of < 0.71 for boys and r > 0.93 with a SEE of < 0.71 for girls (Table 2.11).[311] These results suggest that the measurement of body water compartments with multifrequency bioelectrical impedance in children 9 to 11 years old is an accurate and valid procedure, and the values are comparable to those reported by other studies in children and adults of bioelectrical impedance measurements.[42,250,301,687] The major advantage for the bioelectrical impedance method is that the measurement error among testers is minimized.[283,388] This contrasts with the anthropometric measurements, where it has been shown that tester differences can introduce substantial error in the measurement of skinfolds.[379] However, the use of the bioelectrical impedance method to assess changes in an individual over time must ideally control for biological and environmental variables such as hydration status, timing, content of last ingested meal, and skin temperature.[283]

A new computerized optical system (LIPOMETER) permits a non-invasive, quick, precise, and safe determination of the thickness of subcutaneous adipose tissue (SAT) at specific body sites.[437] The LIPOMETER measures the SAT layer thickness of 15 specified body sites, which are well distributed over the whole body from neck to calf. This SAT topography rebuilds the fat distribution pattern of a subject precisely.[437] Measurements are performed on the right side of the body, one after the other, while subjects are standing. The complete SAT topography determination of one subject lasts approximately 2 minutes. The sensor head of the LIPOMETER consists of a set of light-emitting diodes (wave length 660 nm, light intensity 3000 mcd) as light sources and a photodetector as a sensor. This sensor head is held perpendicular to the measurement site. Because of the relatively large rectangular dimensions of the sensor head (38 × 25 mm), there is no particular influence from the pressure applied during the measurement. The light-emitting diodes illuminate the SAT layer through the skin, forming certain geometrical patterns that vary in succession. The photodiode measures the corresponding light intensities backscattered in the SAT, showing principal proportions depending on the thickness of the measured SAT layer. The light signals are amplified, digitized, and stored in a computer via a special interface card. Shielding from outside light sources is not necessary due to the special interface card.[437]

Calibration and evaluation of the LIPOMETER was done using the CT as the reference method.[437] CT scans were performed at the same 15 body sites as LIPOMETER measurements. The paired values of the LIPOMETER and the CT reference were strongly correlated (r = 0.98 for n = 158).[437] The LIPOMETER has been used to measure fat mass in prepubertal children in our laboratory and has shown close correlations with the fat mass measured by bioelectrical impedance analysis.[318] Thus, this new computerized optical system could be used to assess body composition in prepubertal children.

2.5.3 Changes in body composition in prepubertal children

Relative body fat continues to increase to a maximum value during the first 6 months of postnatal life, then falls to a nadir of about 13% in boys and 16% in girls in late childhood.[212] Many studies indicate that rankings of fat demonstrate relative stability during prepubertal years.[115,212,515,521] The percent of body fat calculated from the two-component model increases slightly in early adolescence but decreases by an average of 1.1% per year from 10 to 18 years in males.[115,515] In females, there is little change during the same age range.[115,515] However, these conclusions may need revision when multi-component models are used.[515]

Total body fat calculated from body density does not change in boys before puberty, but there is an average increase of 1.1 kg per year in girls.[115,515] In addition, Chumlea et al.[115] reported an average increase of 4.4 kg fat-free mass per year in boys from 10 to 18 years, with only slight changes in girls. In a sample of 102 girls, Young et al.[690] demonstrated that, from the ages of 9 and 10 years to 16 years, the skinfold thicknesses increased by 51% and body density decreased by 0.7%. Using DEXA or bioelectrical impedance analysis in healthy Dutch children and adolescents (aged 4 to 20 years), percent of body fat was higher in girls than in boys at all ages.[78]

The results from the 1981 Canada Fitness Survey[207] demonstrate that the prepubertal years in boys and girls are associated with an increase in mean skinfold thicknesses measured at different sites of the body with advancing age groups. At the age of 11, the sum of triceps, biceps, subscapular, iliac crest, and medial calf skinfolds remains at the same level in boys, while it continues to grow in girls.[207] A pattern of relatively greater deposition of adipose tissue on trunk sites relative to limb sites with advancing age in both boys and girls has been shown, with girls showing a relatively greater limb-to-trunk deposition than boys.[207,465]

Age-related changes in fat mass are generally considered to become manifest after puberty in boys and girls. While boys and girls have similar average stature, body mass, and BMI before puberty,[582] percent of body fat seems to be higher in girls than in boys at all ages when using more detailed body composition analysis. Thus, it is interesting to note that sexual dimorphism in body composition is present in early life, well in advance of mature gonadal function.

2.6   Tracking anthropometric parameters and body composition

Maturation is the process that leads to the achievement of adult maturity. Maturation occurs in all body systems, organs, and tissues. For example, skeletal maturation refers to the radiographically visible changes in the skeleton during growth. Assessments of skeletal maturity (skeletal ages) are associated with body size and shape, the percentage of adult stature achieved, body composition measures such as percent of body fat, bone diameters, the timing of pubescence, and the age at which adult stature is reached.[407,515] Skeletal maturation is thought to be the best method for assessing maturity status and is the only method that spans the entire growth period from birth to adulthood.[407]

Tracking refers to the maintenance of relative rank or position within a group over time. This can lead, for example, to the identification of childhood antecedents of adult obesity and assist in the planning of effective preventive strategies.[43,78,110,377,515] The easiest measure of tracking is an age-to-age correlation.[43,47,78,90,109,110,119,377,405,467,515,516,518,519] Minimally, longitudinal observations of the same individual at two points in time are necessary. Research suggests that the longitudinal method is the only approach that gives a complete description of the growth phenomenon.[153,341,519] Longitudinal data provide an opportunity to describe variation in intensity, velocity, and timing of individual patterns of growth. Correlations (Pearson or rank order) are most often used between the repeated measurements to estimate the tracking of different parameters of growth.[43,55,78,327,412,516,668] Correlations less than 0.30 are considered low, and those between 0.30 and 0.60 are moderate. A stable characteristic is defined as one that exhibits a correlation of greater than 0.50 for two measures obtained at least one year apart.[71] In general, the closer the time span between measurements, the higher the correlation. As the time span increases between the measurements, interage correlations generally decline.[405,518] Factors such as age at first measurement, short-term biological variation, significant environmental change, and measurement variability also influence interage correlations.[405,518]

Methods such as regression modeling techniques have been recommended to track different parameters of growth in children.[641] This method appears to be more sensitive than age-to-age correlation analysis. Traditional tracking analysis has not used the influence of potential confounding factors. Such factors as time, gender, biological age, etc., cannot be taken into account; and, when the longitudinal study has more than two repeated measurements, not all longitudinal data are used to calculate tracking. The Amsterdam Growth and Health Study results have been calculated using a more complicated method, multiple logistic regression analysis.[326] The only negative aspect of this statistical method is that the comparison of findings with previous studies may be more difficult.

Longitudinal data on growth and maturation of prepubertal children active in sports are very limited. These data are essential to assess issues related to potential effects of training for growth and maturation. However, the results of the longitudinal investigation on young Polish athletes are available.[410] In total, 78 boys and 40 girls (enrolled in sports schools in Warsaw, and aged 11 to 14 years) were measured once a year. Statures and body masses of boys and girls enrolled in sports schools were, on average, larger than Warsaw reference data. Estimated velocities of growth in stature and body mass in prepubertal children active in sports indicated earlier maturation compared to reference values in boys, while corresponding values for girls approximated reference medians. On average, boys in sports schools enter each stage of genital and pubic hair development earlier than nonathletes, while girls in sport schools approximated the means for nonathletes. However, the estimated intervals between stages did not differ between athletes and nonathletes of both stages.[410]

The degree to which body fat tracks from infancy to adulthood is important to understand. Childhood obesity is considered one of the most difficult problems in pediatrics[337] and the main nutritional problem in industrialized society.[63,92,352] The relative amount of body fat has important health and health-related fitness implications for children.[43,78,287,309,376,515,634] Research has demonstrated that fatter children have a stronger tendency to be obese in adult life.[279,515,634] Boys with > 25% relative body fat and girls with > 30% relative body fat have a greater risk for developing cardiovascular disease since they have higher systolic and diastolic blood pressure, total cholesterol, and lipoprotein cholesterol ratios.[634,679] Thus, the assessment of obesity in children is important in the early diagnosis and prevention of conditions that are associated in adulthood with hypertension and cardiovascular disease.

Skinfolds and BMI index are the methods most widely used to follow the extent of obesity tracking over several years.[63,186,228,377] For example, a 20-year follow-up study by Garn and Lavelle[228] suggested that obesity does not track strongly. In contrast, Epstein et al.[186] found in a 10-year follow-up study that obesity tracks strongly from infancy to childhood to adulthood.

Research has shown that obesity tracks in prepubertal children only to a limited extent, and the majority of obese children became nonobese by adulthood.[228,377] Because many children who are classified as obese are not very fat in terms of an absolute level of fat content, part of the explanation for low tracking comes from the lack of obese children in terms of percent of body fat.[377] For example, the 85th percentile for 6- to 8-year-old children corresponds to the adult fat content of 17% in males and 22% in females.[382] Thus, when children only above 25% body fat for boys and 32% for girls are identified, the degree of tracking in prepubertal children may be higher.[377] Prepubertal boys above the 85th percentile are estimated to have above 17% body fat, and those above the 95th percentile have more than 23% body fat. For prepubertal girls, the corresponding values are 22% and 27%.[382] Childhood obesity is defined as more than 120% of the ideal body mass in relation to stature-derived age.[146]

Evidence suggests two general trends in body fat tracking. First, subcutaneous fat does not track well from birth to about 6 years of age (subcutaneous fat is very unstable during infancy and early childhood).[43] Second, when individuals at specific positions within a group are considered (the fattest or the leanest at a given age), the fattest children after 6 years of age have a higher risk of remaining fat through childhood and into adulthood.[78] Thus, the risk of excess fat appears to be greater for those who have thicker subcutaneous fat measurements during childhood.

Stature tracks at different degrees from infancy to adulthood.[521] The correlation is weak during early childhood but rises steeply until 5 years of age. It decreases between the ages of 11 and 14 years, and then rises to match the 5-year level at the age of 15.[521] According to Rolland-Cachera,[521] adult stature in males is better predicted by leg than trunk length. In girls, where the prediction is better, the upper and lower body segments appear to be equally predictive.[521]

Inter-age correlations have also been used to determine the stability of the somatotype components during growth.[47,109,119,467] The values for the different somatotype components in adjacent years are generally relatively high, ranging from r = 0.65 to r = 0.99 for endomorphy, r = 0.61 to r = 0.99 for mesomorphy, and r = 0.83 to r = 0.99 for ectomorphy.[47,119,467] However, these correlations drop as the span between years becomes greater. The ability to predict ratings that are three or more years apart is generally poor.[47,119] Carter et al.[109] reported partial correlations between ages for each somatotype component with the other two held constant for three or more years apart to be moderate ($r^2 < 0.35$).

2.7   General considerations

Growth and anthropometric development in prepubertal children is characterized by different changes in individual anthropometric variables. The average stature and body mass in boys and girls increases approximately 5.5 cm and 3 kg, per year respectively, up to the age of puberty. Body fat remains the same in boys and slightly increases in girls before puberty. The somatotypes of boys progress from endo-mesomorphy toward balanced mesomorphy up to 6 years of age, followed by a decrease in mesomorphy and increase in ectomorphy before puberty. In girls, the somatotypes change from endo-mesomorphy toward central somatotypes up to the age of 6. While in prepubertal stages, girls tend toward decreasing mesomorphy, followed by an increase in endomorphy. The somatic growth of prepubertal children tracks on a moderate level only since individual factors such as the rate of growth and maturation influence the tracking of somatic growth.

Differences exist in the anthropometric characteristics between boys and girls through to the age of puberty. The subcutaneous fat is more pronounced in prepubertal girls since they have higher values for skinfolds in comparison with prepubertal boys. The values of different circumferences suggest that the relative muscularity of the lower body is developed in a similar manner in prepubertal boys and girls since the gluteal, thigh, and calf circumferences are not different between sexes. However, sex-linked differences in circumferences of the upper body region are evident in prepubertal children. The robusticity of the skeleton is also more developed in prepubertal boys as they have higher values for the dimensions of the upper and lower extremities as well as on the trunk regions in comparison with prepubertal girls.

Prepubertal children are chemically immature. Prior to sexual maturation, children have more water and less bone mineral content than adults. The density of the fat-free mass changes from prepubertal stages to adulthood. Some body composition methods and prediction equations assume that the individuals measured differ from each other only in the amount of fat, while the density of fat-free mass is the same for all individuals. This is not the case for prepubertal children and, thus, the estimation of body composition in prepubertal children demands a carefully chosen measurement procedure.

Tracking refers to the maintenance of rank order within a group of subjects over time. The degree to which body fat tracks from infancy to adulthood is important to understand as childhood obesity is considered to be one of the most difficult problems in pediatrics. The evidence based on research data suggests that subcutaneous fat does not track well from birth to about 6 years of age. However, when individuals at specific positions within a group are considered (the fattest or the leanest at a given age), the fattest children after 6 years of age have a higher risk of remaining fat through childhood and into adulthood. The assessment of obesity in children is important in the early diagnosis and prevention of conditions that are associated in adulthood with cardiovascular disease.

The assessment of body composition in prepubertal children can be done using several sophisticated techniques. However, in many circumstances it is more desirable to utilize widely available and simple techniques such as anthropometry and bioelectrical impedance analysis — allowing rapid and valid determination of body composition in field settings. Other methods may need special laboratory equipment (deuterium oxide dilution), may be impractical for prepubertal children (underwater weighing), may involve radiation exposure (CT scanning), and may require prohibitively expensive equipment (DEXA).

Monitoring the anthropometric and body composition status of prepubertal children provides information about their current health status as well as future health risks. At present, the most appropriate method for the assessment of body composition in prepubertal children is bioelectrical impedance analysis. This technique is less invasive, requires less technical skill, has both higher inter- and intratester reliability, is faster, and may be easier to administer to young children in comparison with skinfold measurements. More detailed anthropometric measurements are also useful in monitoring the growth of children, although they may require higher technical skills in the tester. Suggested protocols for the assessment of anthropometric parameters and body composition in prepubertal children are presented in Appendices 1 and 2, respectively.

chapter three

Physical activities of prepubertal children

3.1   Introduction

Physical activity can certainly be considered a health-related behavior in adults. Numerous health benefits of regular physical activities of various intensities have been extensively documented.[31,86,138,219,441,463,479,531] Epidemiological studies have shown that regular physical activity in adults is associated with reduced risk of morbidity and mortality from several chronic diseases, particularly coronary heart disease.[441,463,664] The prevalence of chronic diseases is negatively correlated with physical activity.[67,461] This correlation cannot be found in children since these types of diseases rarely affect children.[68] However, certain risk factors associated with chronic disease have been observed in children.[219,231,681] Therefore, risk factors for chronic diseases are more typically used to assess the relationship between physical activity and health in children.[505]

Since regular physical activity is associated with health status in adults, it seems reasonable to propose that children should be physically active during childhood so that they may carry this behavior into adulthood. Several studies[153,462,478,548,588,622,623] provide evidence that childhood physical activity habits may determine adult levels of physical activity. Thus, the main question is, how physically active are children today? Although the answer to this question may seem easy to obtain, it is complicated by the fact that reliable and valid methods used to assess physical activity in children are only beginning to emerge,[219] despite the application of at least 30 different methods.[218,557]

Physical activity, exercise, and physical fitness are sometimes used interchangeably. However, they are distinct concepts. Physical activity has been defined as any bodily movement produced by skeletal muscles that results in energy expenditure above the resting level.[111,112] Physical activity has been described to have four basic dimensions:[438,551]

- Frequency — sessions per day or week
- Intensity — the rate of energy expenditure, corrected for body mass, often indicated by kilocalories expended per minute or by multiples of resting metabolic rate; also can be reflected by the percent of maximum heart rate or percent of VO_{2max}
- Duration — minutes or hours per day or week
- Mode — muscle groups involved

The assessment of physical activity can be expressed as the amount of total work performed (in watts), as the time period of physical activity (in minutes, hours), as units of movements (in counts) and/or as a numerical score obtained from the responses to a specific questionnaire.[438] Furthermore, physical activity is often expressed in terms of energy expenditure.[438,551]

Exercise is considered to be a subcategory of physical activity and is defined by Caspersen et al.[112] as "physical activity that is planned, structured and repetitive bodily movement done to improve or maintain one or more components of physical fitness." Physical fitness is a set of personal characteristics achieved from regular physical activity. These characteristics include cardiorespiratory endurance, muscular endurance, muscular strength, body composition, and flexibility.[111,231,551]

Biological differences exist in physical activity patterns between children and adults. Rowland[531] emphasized that children are inherently active primarily because it is physical movement that provides them with the necessary information required by the central nervous system for stimulation. Children have an inherent biological need to be physically active. In contrast, adults achieve arousal of the central nervous system in a variety of non-loco-motor activities such as reading and writing. As a rule, the duration of moderate to vigorous activities is relatively short in all children, especially in preschool children. Bailey et al.[24] found that nearly all bouts of vigorous activity (95%) lasted less than 15 seconds and only 0.1% of the bouts were longer than a minute. On the other hand, children do not remain inactive for extended periods of time. These findings clearly document the transitory nature of children's physical activities that are probably necessary for normal growth and development. More detailed cognitive and behavioral differences between adults and children that should be considered when studying or promoting physical activity were recently presented by the National Association for Sport and Physical Education.[138]

Physical activity is an essential prerequisite of health since human beings are born active. This applies to all age categories, including prepubertal children. It is generally assumed that the more active people are, the fitter they are. However, it is necessary to distinguish between the terms physical activity and physical fitness,[533] a particularly pertinent distinction in children.

Physical activity is a behavior, whereas physical fitness is an attribute.[140,505,533] Physical fitness is affected by genetic inheritance, maturational status, and activity levels, although the relative contributions of each are unclear.[505,533] Physical fitness levels of children have been measured several times. However, it is still not clear whether physical performance capabilities in any exercise event can be directly related to health outcomes.[140,533] Physical activity is the more pertinent variable to assess regarding health in children since impaired physical activity can cause deviations from normal growth and development of the human organism.[533] Furthermore, the lack of physical activity in children plays a role in the pathogenesis of some diseases manifested later in life.

3.2   Health benefits of physical activity

During the past several decades, interest in the benefits of exercise has prompted increasing research to examine the relationship between regular physical activity and health status. Because most attention has been paid to disease end-points and, therefore, has focused on adults, the childhood physical activity level in relation to adult health status is less well defined. Physical activity affects many systems of the human body and provides numerous health benefits for adults.[67,86,461,479,551,664] Because physical activity provides significant protection from chronic diseases such as cardiovascular diseases and non-insulin-dependent diabetes–mellitus — and because it appears to reduce the risk of osteoporosis and some cancers — there is a substantial interest in beginning the prevention of these adult diseases during the first decades of life through regular physical activity.[551] Health concerns such as adiposity, psychological functioning, immune status, and risk of musculoskeletal injury may be influenced by physical activity in children.[551,556] In addition to disease prevention benefits, physical activity enhances the quality of life, enhances the ability to meet physical and mental working demands, and allows children to engage in leisure activities.[551,556] Children who take physical education classes daily throughout the school year perform better academically, have less absenteeism, and are more disciplined.[22,576] The effects of childhood physical activity on chronic disease in adulthood continue to be controversial because long-term studies have not been done.[261,505,551] Other unknowns are the amounts and types of physical activities during childhood that are appropriate for optimal health.[505]

Disease end-points are commonly used to assess the relationships between physical activity and health status in adults.[505,551] However, disease end-points are inappropriate for children; risk factors for coronary heart diseases, especially for chronic heart disease, are more typically used.[505] Risk factors for coronary heart disease are reportedly evident before adulthood.[48,80,505,545] For example, investigations in the U.K.[80] and in the U.S.[48] indicate that over 69% of 12-year-old children have at least one modifiable risk factor for coronary heart disease. Although the clinical manifestations of such diseases do not usually appear before adulthood, many studies have reported that advanced atherosclerotic lesions are identifiable in children.[505,533,545,557] Fatty streaks have been found in the arteries of some children less than 3 years old and in all children older than 3.[533,545,557] These fatty streaks begin to appear in the coronary arteries by the age of 10 years.[533] There is an increasing concern that sedentary lifestyles in childhood lead to an increased risk of lifestyle-related diseases in adulthood.[505] Evidence suggests that risk factor status for dyslipidemia, obesity, hypertension, and physical fitness tracks into adulthood.[48,505]

Risk factors in childhood are considered significant determinants of coronary heart disease risk status in adulthood.[505,533,543] Physical activity in children is reportedly associated with hypertension[217,551,608] and obesity.[163,558,661] Furthermore, increased physical activity in children may elevate HDL concentration in blood.[367,558] Sallis[543] has suggested that regular physical activity practiced from childhood may reduce the tracking of several risk factors. However, the mechanism by which childhood physical activity influences the tracking of coronary heart disease risk factors is still unknown.[533,543] The most important reason for encouraging physical activity in childhood is the presumed tracking of this activity into adulthood.[153,305,462,478,533,548,588,622]

Table 3.1   Health Benefits of Regular
Physical Activity in Children

Body Composition Development
Obesity prevention
Body fat reduction
Fat-free mass development
Enhanced skeletal development
Musculoskeletal injury prevention
Increased muscle and bone strength

Improved Cardiorespiratory Fitness
Moderated blood pressure
Increased HDL[a] cholesterol
Decreased total and LDL[b] cholesterol
Decreased triglycerides
Lowered risk of developing diabetes

Improved Psychological Health
Depression prevention
Anxiety/stress prevention
Reduced symptoms of depression and anxiety
Increased self-esteem/self-concept

Improved Immune Status

Improved Agility and Functional Independence

[a]HDL — high-density lipoprotein.
[b]LDL — low-density lipoprotein.

The summary of health benefits of regular physical activity in children is presented in Table 3.1. It is clear that regular physical activity plays an important role in a child's health. The promotion of regular physical activity in children should be a priority for all health professionals in order to encourage children to adopt and maintain healthy and active lifestyles. However, we do not know the varying effects of different sports events on health. For example, the effects of soccer in boys and aerobics in girls may have different impacts on the health of children. No research data are available about the influence of so-called modern sports events such as inline skating or martial arts on the health of children. Is it enough for children to walk to school and back?

3.3   Assessment of physical activity

Before the relationship between daily physical activity levels in children and health can be determined, a valid method for the assessment of daily physical activity is needed. Development of effective programs for the promotion of physical activity in children is also important. Lack of these methods and programs has been the major limiting factor in this type of research to date. Ideally, it is desirable to record the normal daily energy expenditure of a child.[533,670] The various methods currently available to assess physical activity vary greatly in their applicability in personal assessment, clinical practice, intervention studies, and epidemiologic research. The results of physical activity monitoring studies can vary greatly depending on how activity is measured and interpreted. The measurement methods used to assess physical activity in children include the following procedures (Table 3.2):[219,427,533,670]

- Direct observation of activity
- Doubly labeled water
- Activity recall or record
- Questionnaires directed to the child, parent, or teacher
- Heart rate monitoring
- Indirect calorimetry
- Mechanical motor sensors

The major shortcoming of all techniques is that the validity, reliability, and objectivity of many of these methods are not yet well established. The matter is complicated in that there is no universally accepted criterion method to validate measures of physical activity and caloric expenditure.[219,427,533]

Table 3.2 Comparison of Methods used in Physical Activity Research in Prepubertal Children

Method	Advantages	Limitations	Objectivity	Recommendations to use
Direct observation	Can access activity patterns	Requires observers; not practical for epidemiological studies	High	Highly recommended
Doubly labeled water	Highly accurate for assessment of energy expenditure; physiological marker	Limited to total energy expenditure; cannot assess activity patterns, intensity, frequency or duration	High	Highly recommended
Activity recall	Practical for epidemiological studies; can assess activity patterns	Memory and recall errors; proxy reports are needed	Low to moderate	Recommended only for large-scale studies
Activity diary	Practical for epidemiological studies; can assess activity patterns	Errors in estimating activity patterns; proxy reports are needed	Low to moderate	Recommended only for large scale studies
Heart rate monitoring	Physiological marker	Influenced by other factors; limited to moderate to vigorous physical activity (> 140 beats per minute)	Moderate to high	Recommended
Indirect calorimetry	Physiological marker	Must wear a mask during the measurement, which could influence activity	Moderate to high	Recommended
Accelerometer	Energy expenditure and activity accurately predicted	Less sensitive for low-intensity sedentary activities	High	Highly recommended

The procedure that generally serves as the criterion measure of physical activity for children is direct observation,[24,219,334,427,489] wherein the children are either viewed or videotaped for a certain period of time in their normal environments. Observation allows for several different dimensions of physical activity (type, intensity, and duration) to be simultaneously recorded.[24,219,427,489] For greatest accuracy, the sampling interval should be sensitive to brief periods of physical activity (3 to 10 seconds).[24,219,427,489] This is the major concern when monitoring children, whose physical activity patterns are constantly changing.[334] A continuous scoring system designed specifically for children successfully distinguishes eight different physical activities of various intensities as validated by indirect calorimetry.[427,489] The outcome measure is either in kilocalories or some nominal score that is directly related to energy expenditure. The estimated energy expenditure is computed by using the lists of energy costs of various activities and multipling by the time allocation.[219,438] However, the translation of physical activities to caloric expenditure usually involves the use of adult energy expenditure values, which may underestimate the level of energy expenditure in children.[219,547] Direct observation is very time consuming and expensive since it is necessary to employ one observer for each subject.[219,547] Nevertheless, direct observation is a valuable procedure for the validation of other less time-consuming measurement methods.[219,335]

The doubly labeled water technique has also been used as a field-based validation measure of energy expenditure in everyday situations.[219,427,438,556,566] This method measures the disappearance rate of a labeled isotope ($^{2}H_2^{18}O$) from urine samples to estimate carbon dioxide production over a period of several days. This method is applicable to children since the disturbance is limited to drinking water and collecting urine samples.[556] Evidence suggests that doubly labeled water estimates of energy expenditure are most accurate when monitoring is done over a 6- to 14-day period.[427,556,566] Results of several validation investigations have demonstrated that the doubly labeled water technique could overestimate energy expenditure by 2 to 8%.[427,556,566] However, the high cost of labeled isotopes limits the use of this method to relatively small groups of subjects. Another limitation of this technique is that information is limited to total energy expenditure, with no frequency, intensity, or duration information.[219,427,438] Despite these limitations, the doubly labeled water method is a technique that will continue to be used for the validation of other less direct measurements of physical activity.

Questionnaires (self-reports) are commonly used in large groups of subjects in epidemiological studies.[111,219,235,305,343,427,544,556,670] Four major types of self-reports are used in children according to Sallis:[544]

1. Self-administered recall — children report their own activities on a preprinted form; responses may be open- or close-ended.
2. Interview-administered recall — interviewers administer a structured interview with a child in a one-on-one session.

3. Diary — children code physical activities throughout the day in a diary form.
4. Proxy reports — parents or teachers report the activity of a child using any of the three previously listed formats; some measures use parent reports to supplement child reports.

The advantages of self-report methods are: (1) they are unobtrusive; (2) they are relatively easy to administer and score; (3) they are relatively inexpensive; and (4) many variables can be assessed with a single instrument (leisure vs. occupational activity; duration, intensity and frequency of activity; and estimated caloric expenditure).[30,427,544,670] However, the self-report procedure is not recommended for prepubertal children since they cannot be expected to recall or record their physical activity levels with great accuracy. The errors introduced may make it insensitive to individual differences in the levels of physical activity.[301,343,472,556,557,635]

The time is especially difficult to recall. Children generally overestimate the time spent in vigorous activities like playing outside; and they underestimate time spent in regular activities such as going to school or eating.[556] Proxy reports can be used to assess activities of children too young to report their own behaviors.[343,544] The use of questionnaires in prepubertal children is limited by the: (1) sporadic nature of children's activities;[219,335] (2) inability of the child, parent, or teacher to accurately recall type, intensity, and frequency of physical activity;[219,343,472,544,557,635] and (3) errors associated with converting children's physical activities to caloric expenditure equivalents.[219,633] Thus, the more diffused, disorganized, and spontaneous the activity — which is typical for children — the more difficult it is to recall. Furthermore, Sallis[544] has reported only moderate relationships between various self-report forms and other objective criteria because children cannot provide accurate self-report information about their activity patterns.

Heart rate monitoring has also been used as a valid and practical indicator of physical activity in children.[219,232,328,372,453,533,556,599,670] The linear relationship between heart rate and oxygen consumption or energy expenditure during exercise is well established.[219,372,427,533,556] Children who spend longer periods of time in higher heart rate ranges are generally more active than those children whose heart rates are in lower ranges.[219,328,453,599] However, heart rate may also be elevated due to emotional stress, changes in posture, temperature, type of muscle contraction (static or dynamic), and the muscle mass involved.[267,359,372,533] Nevertheless, emotional stress is unlikely to cause a prolonged elevation in heart rate such as that caused by prolonged vigorous physical activity.[464,549] The return of heart rate to baseline may also lag behind the return of oxygen consumption to baseline.[557] Fitness level affects heart rate responses, with the more fit child having a lower heart rate at a given level of energy expenditure.[219,533,558] Saris et al.[558] observed that children with higher physical performance capabilities who were also highly active (according to questionnaires completed by parents) had lower mean daily heart rates than children in a low physical performance and low activity group. The more fit and active children spent longer periods at lower heart rate levels than less active children. The lower heart rate levels were attributed to the fitter children having higher stroke volumes and accordingly lower heart rates for a given physical activity.[558] Mean daily heart rates may therefore be more representative of children's physical fitness than their activity levels.[558]

All of these factors may make the conversion of heart rate to energy expenditure inaccurate.[219,505,533,670] Heart rate monitoring should be primarily considered as a tool for the assessment of moderate to vigorous physical activity.[505,533] Heart rates below 120 beats per minute are not normally valid estimates of physical activity.[505,533] Furthermore, heart rate monitoring is not suitable for gaining a picture of all the different daily activities of children since over 75% of the day may be spent at heart rates less than 120 beats per minute.[218,232,533] However, heart rate monitoring has successfully been used to distinguish activity patterns in children, providing an indication of the intensity, duration, and frequency of physical activity.[232,427,549] For example, Gilliam et al.[232] monitored heart rate for 12 hours to determine the amount of time children spent in physical activity strenuous enough to promote cardiovascular fitness (>160 beats per minute) during one day without school. Heart rate monitoring could be an appropriate measure to assess hard (>140 beats per minute) and very hard (>160 beats per minute) physical activity of children.[549] However, the limitations of this tool bring into question the suitability of heart rate monitoring for the validation of other methods that may be used in epidemiological studies in children.

Indirect calorimetry is one of the most accurate techniques to assess daily physical activity and could be used to validate physical activity monitors in field settings where energy expenditure is required as a criterion measure.[426,530,556] With indirect calorimetry, energy expenditure is determined from oxygen consumption and carbon dioxide production.[426,530,556] Children must wear masks during the whole day, which could potentially influence daily activities.[530]

The use of mechanical or electrical motion sensors in children assumes that the movement of the limbs or the whole body reflects total energy expenditure.[219,427,670] The advantage of motor sensors in the assessment of physical activity in children is that they are less costly and time consuming in comparison with many other methods. Furthermore, since the motion sensors directly assess movement, they are likely to be more sensitive to variation in physical activity than, for example, questionnaires or activity recall. Technological advances have led to the development of small, lightweight instruments that can store data over many days and permit analysis of data within discrete intervals of time. Different motion sensors offer great flexibility to the user. The accelerometer, actometer, large-scale integrated motion sensor, and pedometer are examples of motion sensors that have been used in children.[219,533,670]

The use of accelerometry to measure physical activity is based on the assumption that accelerations of the limbs and torso closely reflect energy cost.[219,427,438,546,556,670] Caltrac™ is the most prominent accelerometer employed in physical activity research in children since it is sensitive to the amount and intensity of movement in the vertical plane. Vertical accelerations result in twisting an internal ceramic piezoelectric transducer; and the amount of twisting is proportional to the size of the acceleration.[219,427] The Caltrac accelerometer functions either as an activity monitor, which provides activity counts based on vertical acceleration as the individual moves about, or as a calorie counter, in which the acceleration units are used in conjunction with body size (body stature and mass), age, and sex to estimate energy expenditure. Laboratory and field investigations to validate the Caltrac accelerometer have been conducted with children. For example, Sallis et al.[546] studied the validity of the Caltrac accelerometer by comparing Caltrac readings to energy expenditure measured using indirect calorimetry during exercise on a treadmill. The correlation coefficient between energy expenditure measured by the Caltrac accelerometer and indirect calorimetry was $r = 0.82$ in children 8 to 13 years old.[546] However, the Caltrac accelerometer does not reflect caloric expenditure of all types of physical activities because it is calibrated to the single activity of treadmill walking.[427] The Caltrac accelerometer should be used in combination with other measures of physical activity.[427,546]

The new version of the Caltrac monitor — the TriTrac R3D™ activity monitor — is based on the same accelerometry principles as the Caltrac monitor but can measure quantity and intensity of the movement in three dimensions rather than in one. It also features an internal clock mechanism that allows activity to be assessed on a minute-by-minute basis. Furthermore, it features solid-state construction and a serial interface that allows it to download data into a computer.[669,670] A triaxial accelerometer provides more accurate estimates of low levels of energy expenditure that are not well represented by movement in the vertical plane.[125,416,670]

Welk and Corbin[669] studied the validity of the triaxial accelerometer as a field measure of physical activity in prepubertal children on three different school days with heart rate monitors.[669] The correlations between average vector magnitude from a triaxial accelerometer and average heart rate (corrected for resting values) were $r = 0.51$ to $r = 0.89$ (mean $r = 0.57$)[669] in comparison with correlations of $r = 0.51$ to $r = 0.69$ (mean $r = 0.57$) obtained by Janz[302] with a uniaxial accelerometer. Both studies included three days of monitoring. Welk and Corbin[669] spread three days over an eight-month period, whereas Janz[302] assessed activity over three consecutive days. Although the sample sizes were similar in the two studies, a much greater age range existed in the study by Janz[302] (7 to 15 years) compared with the study by Welk and Corbin[669] (9 to 11 years). This disparity limited the comparability between the two studies. However, by taking advantage of the minute-by-minute timing capability of the triaxial accelerometer and the heart rate monitor, Welk and Corbin[669] discovered that the correlations between these instruments were highest during free play situations (recess, after school) and were lower when activity was more limited (class time) or structured (physical education). Thus, independent of the possible benefits of a three-dimensional assessment, the primary advantage of the triaxial TriTrac R3D accelerometer may lie in its ability to segment activity by time.[669]

Studies in another laboratory demonstrated that TriTrac R3D correlates more highly with scaled oxygen consumption ($r = 0.91$) than heart rate ($r = 0.79$) across a variety of activities in 30 children aged 8 to 10 years.[191,192] This suggests that there are more appropriate criterion measures than heart rate monitoring to validate the TriTrac R3D accelerometer.[191,192] Nevertheless, this triaxial device provides a relatively accurate estimate of energy expenditure in free living environments and has potential use for validating other less sensitive measurements of physical activity in children.[191,192,533,670]

Recently, Trost et al.[637] assessed the validity and inter-instrument reliability of the new CSA 7164 Activity Monitor (Shalimar, Florida, U.S.) in children aged 10 to 14 years. This new CSA monitor is a uniaxial accelerometer designed to detect vertical acceleration ranging in magnitude from 0.05 to 2.00 G with frequency responses of 0.25 to 2.50 Hz. Activity counts were strongly correlated with energy expenditure during treadmill walking and running ($r = 0.86$). The intraclass correlation for two CSA 7164 monitors worn simultaneously was $r = 0.87$, indicating a strong degree of inter-instrument reliability. Unfortunately, no data are available about the relationships among Caltrac, TriTrac R3D, and CSA 7164 in children. Trost et al.[637] indicated that a seven-day monitoring protocol provided reliable estimates of normal physical activities and accounted for important differences in weekend vs. weekday activities in large groups of children. Differences in activity patterns within a given day were discerned using the CSA 7164 uniaxial accelerometer.[637]

Eston et al.[192] tested the validity of uniaxial pedometers, triaxial accelerometers, and heart rate monitors for predicting energy expenditure in children while walking and running on a treadmill, during two brief recreational activities, and during one sedentary activity. Each of the measures was significantly correlated with energy expenditure, and the accelerometer was the best predictor of energy expenditure. Pedometer and heart rate monitors were similarly related to energy expenditure.[192] Important differences may exist in the sensitivity of the pedometer to activity counts during moderate to intense physically active play vs. low-intensity, sedentary activities; the pedometer is less sensitive to low-intensity, sedentary activities.[192]

Kilanowski et al.[332] compared the activity measurements during recreational physical activities and low-intensity classroom activities in the natural environments of children 7 to 12 years old using an electronic pedometer, a triaxial accelerometer, and behavioral observation. The pedometer, accelerometer, and behavioral observation measures were highly correlated for combined activities and recreational activities, equaling or exceeding $r = 0.95$ ($p < 0.001$). However, correlations between the pedometer and accelerometer were significantly lower during low-intensity classroom activities vs. recreational activities (r = 0.98 vs. r =0.50; p <0.05).[332] The activity patterns of most children include brief bursts of moderate- to high-intensity physical activities combined with periods of low-intensity activities.[232,332] Pedometers are designed to register activities only in vertical directions — not back-and-forward or side-to-side movements. However, low-intensity classroom and other sedentary behaviors involve sitting, with little movement to vertical direction. Uniaxial pedometers are less sensitive to low-intensity physical activities that do not involve vertical movement by children.[192,332] Consequently, triaxial accelerometers outperform uniaxial devices in predicting energy expenditure during sedentary activities such as writing, sitting, or standing.

There are other limitations to the use of uniaxial pedometers vs. triaxial accelerometers:[332,670] (1) pedometers provide only estimates of cumulative activity and do not record or store activity data by time; and (2) it is not possible to determine activity parameters (duration of the exercise, intensity of the exercise, or number of discrete exercise bouts per day). The one-dimensional nature of pedometers is a particular problem for children, who engage in play activities that involve greater diversity of movement than many repetitive aerobic activities of adults.[332,670] Pedometers are better suited to assess physical activities of higher rather than lower intensity; they are the choice if the goal of the study is to assess differences in physical activities among moderate- to high-intensity behaviors.

The choice of methods for the estimation of physical activity and energy expenditure in children in free living situations should consider the objectiveness, cost, validity, and reliability of the methods. Despite the considerable research in this area, more validation studies are necessary. The activity monitors provide valid measures of physical activity, but they bring questionable estimates of energy expenditure.[670] Further research must also focus on standardizing physical activity assessment procedures. Some of the objective methods are very promising for use in children, but further work is needed to determine the amount of day-to-day variation in physical activity and the best way to attach the devices to minimize malfunction.[219,670] Because each of the methods has limitations, use a combination of methods to assess the physical activity in prepubertal children. The most suitable methods of measuring physical activity in prepubertal children are summarized in Appendix 3.

3.4   Physical activity guidelines

We do not know the amount and type of physical activities during childhood that are conducive to optimal health maintenance.[305,505,551] Activity levels known to provide health benefits to adults are also generally appropriate for children.[505,551] There are two health-related rationales for children to be physically active:[505,551] (1) to promote physical and psychological health and wellbeing during childhood; and (2) to promote physical activity to enhance future health by increasing the probability of remaining active during adulthood.

The traditional aim of an exercise program is to improve cardiovascular fitness.[10,138,505] An activity of 20 to 60 minutes' duration, using large muscle groups, performed three to five days per week, at an intensity of 60 to 90% of theoretical maximum heart rate (or 50 to 80% of VO_{2max}) is recommended.[10] However, cardiorespiratory fitness and cardiovascular health are not synonymous terms. Physical training adaptations may not be directly related to good health or disease prevention[140,505] since health benefits may accrue at intensities of physical activity below those necessary for marked improvement in physical fitness.[505] This distinction may be crucial since less than 10% of adults are engaged in cardiovascular fitness training.[505,607] Vigorous physical activities are an excellent way to increase activity for a minority of individuals; but there are many other ways to obtain health benefits from being physically active. Unfortunately, most investigators conducting research on activity levels of both children and adults have adopted cardiovascular fitness training criteria as the activity level necessary to confer health benefits.[505] From a behavioral perspective, physical activity has to be seen by children as an achievable and positive experience.[103] Adult fitness training guidelines, emphasizing continuous bouts of vigorous exercise, do not fulfill these criteria for children.[103]

There is no universally accepted consensus regarding physical activity patterns in prepubertal children. Any increase in physical activity under safe conditions furthers the development of the child.[466] Many researchers and organizations have proposed various physical activity criteria for children (Table 3.3)[10,66,266,525,536,574,585,622] However, a stronger scientific base on which to create physical activity guidelines for prepubertal children is needed. As emphasized by Riddoch and Boreham,[505] the use of different criteria could lead to markedly different conclusions about whether children are sufficiently active. Early childhood is a critical period in forming adult physical activity habits. Different kinds of physical activities in children should provide a base for positive attitudes, knowledge, and skills to carry over into adulthood.[266] These activities include vigorous activities such as brisk walking, jogging, swimming, bicycling, and aerobic dance.[606] These activities do not carry over into adulthood behavior when children do not engage in them with regularity, intensity, and duration.[543]

Table 3.3   Suggested Physical Activity Criteria for Children

Reference	Country	Year	Suggested physical activity levels
Blair et al.[66]	U.S.	1989	Minimum exercise energy expenditure of $3 \text{ kcal} \cdot \text{kg}^{-1} \cdot \text{day}^{-1}$
Pyke[536]	Australia	1987	Vigorous physical activity, 3-4 times per week, at least 30 minutes per session
Shephard[574]	Canada	1986	Three hours per week, ~25 minutes per day, 4 METs[a] intensity
American College of Sports Medicine[10]	U.S.	1991	Three times per week, 20 minutes per session, intensity at or above 60% maximal oxygen consumption
Ross and Gilbert[525]	U.S.	1985	Minimum of 3 times per week, 20 minutes per session, intensity at 60% of cardiorespiratory capacity using large muscle groups
Telama et al.[622]	Finland	1994	At least 30 minutes activity per day, every day
Haskell et al.[266]	U.S.	1985	At least 30 minutes activity per day, every day, using large muscle groups
Silvennoinen[585]	Finland	1984	At least 2 times per week, activity causing continuous breathlessness and abundant sweating

[a]METs — metabolic equivalents.

The first physical activity guidelines for children were similar to those recommended for adults.[10,525] More recently, the recommendations have been revised to accommodate the differences between children and adults. Relatively low target heart rates are recommended for children. Twenty years ago Gilliam et al.[232] recommended a heart rate of 160 beats per minute. However, Simons-Morton et al.[588] indicated that a heart rate of 140 beats per minute is sufficient to define the threshold for activity in children.

The relatively high fitness levels in children may result from large volumes of sporadic physical activity performed at low intensity, which does not conform to any physical activity guidelines.[505,550] Chasing, climbing, wrestling, and all playground games contribute to the improvement of health-related physical fitness.[505] It is a great challenge for researchers to assess the variety of activities in which children engage, and the methodological problems are also considerable.[550]

A more precise identification of the amounts and types of physical activities that are appropriate for the health and well being of children is necessary. These physical activity guidelines should be based on longitudinal studies of children that monitor health and activity behavior. The guidelines presented at the 1993 International Consensus Conference on Physical Activity Guidelines for Adolescents are appropriate for prepubertal children as well. Adolescence was defined as ages 11 through 21 years. The consensus document[551] suggested that all adolescents should be physically active daily, or nearly every day, as part of their lifestyles. They should also engage in three or more sessions per week of activities that last 20 minutes or more and that require moderate to vigorous levels of exertion.[551]

3.5   Physical activity of children in different countries

Little is known about the actual physical activity levels of children, although they are highly and spontaneously active.[505,530,590,636] Simple observation tells us that young children are constantly on the move and have endless energy. However, the use of different criteria has led to different conclusions about whether children are sufficiently active. Some researchers have reported that young children are sufficiently active to achieve a health benefit,[66,505,556] while others report physical activity levels so low that they are detrimental to health.[22,102,231,543] For example, Blair et al.[66] report that about 90% of American children are sufficiently active. In contrast, Sallis[543] describes the activity levels of children as shockingly low. The different findings may be due to the different activity thresholds that have been used.[505] Investigations that use more objective methods report much lower levels of activity, while the use of less stringent health-related thresholds results in higher levels of appropriate activity in children. Thus, the findings are not directly comparable. However, it is possible to develop an appropriate estimate of the proportion of young children who are active in different countries.

The 1981 Canada Fitness Survey involved 2702 boys and 2576 girls 7 years and older.[105] Children's physical activity levels were measured by a self-report technique. The criterion of sufficient physical activity was three hours of moderate activity per week with at least four metabolic equivalents (METs) intensity, which approximated 25 minutes of activity per day. Boys and girls (73% and 70%, respectively) met this criterion. However, when the criterion of participation in vigorous activity for three hours per week was applied, less than 5% of children were sufficiently active.[105] A national survey in Australia[536] asked if children aged 9 to 15 were vigorously active 3 to 4 times per week for at least 30 minutes per session. It was found that 50% of boys and 39% of girls were active at this level. Furthermore, there was no apparent decline with age for either boys or girls.[536] In comparison, a study of 6500 children from the U.K. found that less than 49% of boys and 19% of girls engaged in vigorous activity 3 or more times a week for at least 30 minutes a session.[102]

A national survey in the U.S. included 10,275 boys and girls aged 10 and older.[525] A self-report technique was used. Appropriate physical activity was defined as a minimum of three sessions per week with 20 minutes of exercise per session at an intensity of 60% of cardiorespiratory capacity using large muscle groups. It was found that 61% of boys and 57% of girls engaged in appropriate physical activity year-round. Furthermore, boys spent 114 minutes and girls 103 minutes per day in sports, active games, and exercises outside physical education classes.[525]

The physical activities of Estonian prepubertal boys and girls have been studied in our laboratory,[496-499] with 418 7- to 10-year-old children selected randomly from elementary schools in the city of Tartu. All children were healthy and participated in school physical education lessons twice a week. Physical activity was assessed by a 7-day physical activity recall modeled after Godin and Shephard.[235] Every day during one week, parents were asked to report on how much time (in hours and minutes) their children spent on activities outside of school. Activities were classified as low (3 METs), moderate (5 METs) or vigorous (9 METs). Classification of activities in each category was completed using the Compendium of Physical Activities designed by Ainsworth et al.[3] Examples of physical activities most frequently performed by children were selected (sports, home activities, leisure activities). Special attention was given to moderate to vigorous physical activities such as play games and sports activities.

In a pilot study, the validity of the 7-day physical activity recall used in our investigation was administered using the Caltrac accelerometer as a field measure of physical activity. The correlation coefficients between the Caltrac counts and physical activity recall were $r = 0.34$ to $r = 0.46$ for weekdays and $r = 0.44$ to $r = 0.61$ for weekends. Physical activity scores for 7-, 8-, 9-, and 10-year-old prepubertal boys and girls are presented in Table 3.4. Boys engaged in significantly more moderate to vigorous physical activity in comparison with girls across all age groups. There were no significant differences in 7- and 8-year-old groups; and 9-year-old girls and 10-year-old boys demonstrated significantly higher scores in total weekly physical activity.

Table 3.4 Physical Activity Scores (Hours per Week) in Prepubertal Boys and Girls Assessed by 7-Day Physical Activity Recall

Gender	MVPA[a]	LPA[b]	TPA[c]
7-Year-Olds			
Boys (n = 53)	12.4 ± 3.4	15.1 ± 3.1	27.5 ± 6.4
Girls (n = 48)	8.9 ± 2.5[d]	16.9 ± 4.9[d]	25.8 ± 6.7
8-Year-Olds			
Boys (n = 45)	11.5 ± 3.1	14.3 ± 4.2	25.8 ± 7.9
Girls (n = 54)	8.3 ± 2.9[d]	16.8 ± 5.6[d]	25.1 ± 5.5
9-Year-Olds			
Boys (n = 50)	9.9 ± 2.4	10.5 ± 2.8	20.4 ± 6.8
Girls (n = 57)	6.1 ± 3.1[d]	17.6 ± 4.8[d]	23.7 ± 6.3[d]
10-Year-Olds			
Boys (n = 55)	7.3 ± 2.3	14.3 ± 3.6	21.6 ± 5.9
Girls (n = 56)	4.9 ± 2.7[d]	14.6 ± 4.1	19.5 ± 5.8[d]

[a]MVPA — moderate to vigorous physical activity.
[b]LPA — low physical activity.
[c]TPA — total weekly physical activity.
[d]Significantly different from the corresponding value in boys — $p < 0.05$.

Source: Compiled from Raudsepp, L., and Jürimäe, T., *Biol. Sport,* 13, 297, 1996, and from Raudsepp, L., and Jürimäe, T., *Am. J. Hum. Biol.,* 9, 513, 1997.

In a longitudinal study, Telama et al.[622] investigated physical activity and participation in sports of 3596 young people in Finland. At the beginning of the study the boys and girls were 3, 6, 9, 12, 15, and 18 years old. The measurements were carried out in three-year intervals — 1980, 1983, 1986 and 1989. Physical activity was measured by means of a questionnaire. Results showed that 90% of boys and 85% of girls were engaged in physical activity for periods longer than 30 minutes at least once a week before puberty, when all measurements at all time points were combined. Also, 79.1%, 82.9% and 76.5% of 9-year-old boys engaged at least twice a week in leisure time physical activity in the years 1980, 1983, and 1986, respectively. The percentage of physically active 9-year-old girls during the same years was 64.3%, 67.0%, and 72.1%, respectively.[622]

Physical activity levels in 86 healthy 10-year-old French children were studied using a validated physical activity questionnaire over the past year.[151] The questionnaires were filled in by both parents and children. Total annual physical activity was 14.3 ± 4.8 hours per week. The boys (n = 54) were more active (15.2 ± 4.9 vs. 12.6 ± 3.9 hours per week) than girls (n = 32). Socioeconomic status was not significantly associated with the level of physical activity. However, children from unskilled workers' families (n = 20) tended to be less active (13.1 ± 4.9 hours per week) than children from families of intermediate (n = 33) and executive professions (n = 33), whose levels were 14.7 ± 5.7 and 14.7 ± 4.2 hours per week, respectively.[151]

In a study of 266 British children aged 11 to 16, Armstrong et al.[14] assessed physical activity levels by heart rate monitoring on three school days and one Saturday. They found that 23% of boys and 12% of girls had at least one 20-minute period of elevated heart rate during school days. However, only 4% of boys and less than 1% of girls had three 20-minute periods of elevated heart rate levels over three school days. Girls were less active than same-aged boys, and there was a greater decrease in physical activity participation among older girls. The authors are concerned that sedentary western lifestyles do not involve sufficient high-intensity exercise and periods of sustained heart rate elevations believed necessary to promote physical fitness.[14]

In a study of children from the developing world, three samples of 10- to 13-year-old Nepali boys (n = 67) were studied.[464] The purpose of this study was to compare levels of physical activity of villagers (n = 31), middle-class schoolboys (n = 20), and homeless boys (n = 16). Methods included continuous heart rate monitoring in conjunction with self-reports and direct observation of physical activity. Mean daytime heart rate (100-104 beats per minute across the three groups) and percentage of time spent vigorously active (heart rate higher than 139 beats per minute — 4%) did not differ between samples despite obvious differences in lifestyles. These results show that lifestyle has little impact on the mean daytime heart rates of 10- to 13-year-old boys.[464]

Very similar values were obtained for children in other developing countries, such as Senegal,[162] Colombia,[605] and Bolivia.[328] The heart rate values reported for boys in developing countries were similar to the values obtained in western countries.[14,372] There are remarkable similarities in mean daytime heart rates of prepubertal boys, regardless of country of origin and socioeconomic status. Thus, the main question is whether the similarities in mean daytime heart rate values reflect similarities in the levels of physical activity and fitness — or whether comparisons of activity patterns between children of different countries need to be based on more sensitive techniques.[464] Time spent in vigorous physical activity is different between children from the western world and those from developing countries. While British schoolboys spent 6.2% of their time above the threshold of 139 beats per minute,[14] the times spent above this threshold for Nepali[464] and Senegal[162] boys were 4% and 1.7%, respectively. The lower levels of vigorous physical activity in boys from developing countries reflect a tendency to spend more time working at a moderate pace to avoid physical exhaustion and to be able to sustain agricultural activities throughout the working day.[464]

Standardization of assessment techniques in health-related physical activity research in prepubertal children is urgently needed. It is often difficult to compare and assess the level of physical activity of children from various countries because the methods used and the definitions applied are different. Many investigations are conducted to validate different physical activity assessment procedures but do not contain numerical data about the actual levels of physical activity in children. However, two main conclusions may be drawn from international studies of children's physical activity: (1) substantial proportions of prepubertal children are not physically active; and (2) higher proportions of girls are less physically active than boys. More studies on cross-cultural comparisons between children of different countries should be conducted.

3.6   Tracking physical activity

A small number of studies have examined the tendency of physical activity behavior to track during childhood.[305,405,475,476] The habits and attitudes toward physical activity developed during childhood are assumed to continue into adulthood.[305,405,588,622] Within relatively short time intervals (3 to 5 years), physical activity behavior tracks over time.[71,305,405,475,476,518,559,622] However, over longer periods of follow-up (6 to 10 years), the tendency for physical activity to track declines considerably.[405,622] Physical activity peaks at the age of 12, after which it starts to diminish.[622] Thus, physical activity during childhood is a significant predictor of physical activity in adulthood.[622] Tracking results of children's physical activities vary considerably with respect to lengths of follow-up, populations studied, assessment of physical activities, and analytical methods used to assess tracking.[405,476]

Data from several longitudinal studies show that children at the extremes of physical activity distribution (children with the highest and lowest levels of physical activity) tend to retain their relative rankings with respect to physical activity over time.[305,476,622] For example, a recent longitudinal study of Finnish children reported that participation in organized sports during childhood is a significant predictor of physical activity in adulthood.[552,622] In sports clubs children learn, develop motor skills, and participate in rather intensive physical activities. Intensive childhood experiences prove important in adulthood, and those who have such experiences can return to physical activities and adopt new types of sports more easily.[552,622] In contrast, Pate et al.[476] found that children at the lowest extreme tended to maintain their physical activity status over three years of investigation during the transition from the fifth to the seventh grade. This suggests that intervention programs to increase physical activity in prepubertal children are warranted.[476]

Table 3.5   Interage Correlations for Indicators
of Physical Activity During Childhood

Span (years)	Correlations	Measurement Method	Reference
3.5 – 6.5	0.57	Heart rate monitoring	Pate et al.[475]
4 – 6	0.27	Direct observation	Sallis et al.[552]
9 – 12	0.44	Questionnaire	Telama et al.[622]
10 – 12	0.36	Questionnaire	Pate et al.[476]
10 – 15	0.32	Questionnaire	Janz et al.[305]

Interage correlations for indicators of physical activity during childhood in different studies are summarized in Table 3.5. Heart rate monitors were used to track physical activities over a 3-year period beginning at 3.5 years of age in American children.[475] Physical activity was quantified as the percentage of observed minutes between 3 and 6 p.m. during which heart rate was 50% or more above individual resting heart rate.[475] Sallis et al.[552] estimated physical activity over a 2-year period beginning at 4 years of age by direct observation of 2 days. In the study of Finnish children, the index of physical activity was derived from questionnaires to estimate intensity, duration, and monthly frequency of participation.[622] Pate et al.[476] used previous day physical activity recall to assess 30-minute blocks of vigorous and moderate to vigorous physical activity in rural American children. The Muscatine study by Janz et al.[305] used a 3-day sweat recall to assess vigorous physical activity.

Although different indicators and methods of analysis are used, physical activity tracks at low to moderate levels during childhood. Some tracking suggests that sports activities during childhood form the foundation for activity habits in adulthood. Although tracking inactivity is studied less, research suggests that less active children tend to remain less active in later life.

3.7   The influence of physical activity on anthropometric parameters and motor ability

Habitual physical activity as one of the environmental components of motor development is an important factor in normal growth and development in children.[312,402,407,531] Assessment of physical activity has become more important with growing awareness of the associations among physical activity, health, normal growth, and motor development in children.[31,403,531] Several problems exist in relation to the amounts and types of physical activities that are appropriate for optimal health and motor development in children. In adults, physical fitness is an excellent marker of physical activity,[65,623] while the degree of association, although often significant, is only moderate in children.

Results of several cross-sectional investigations show a significant negative relationship between physical activity and subcutaneous fat.[148,187,251,667] The results from our laboratory agree with these studies (Tables 3.6 and 3.7).[496-499] Moderate negative correlations among the sum of triceps, biceps, subscapular, abdominal and medial calf skinfolds, moderate to vigorous activity, and total weekly physical activities were found across ages in prepubertal boys and girls. Subcutaneous fat is an important factor that affects the level of physical activity in prepubertal children. However, not all studies confirm a significant negative relationship between physical activity and fat in children.[56,70] Energy expenditure has often been presented in absolute energy units without correcting for body mass, particularly in obese children, which could explain the disparate results.[36]

Table 3.6   Zero-Order Correlations Between Physical Activity and Selected Anthropometric Parameters in Prepubertal Boys

Physical Activity	Stature	Body Mass	Sum 5 SF[a]
7-Year-Olds (n = 53)			
MVPA[b]	0.20	-0.31[e]	-0.43[e]
LPA[c]	0.05	-0.12	-0.14
TPA[d]	0.13	-0.24	-0.34[e]
8-Year-Olds (n = 45)			
MVPA	0.17	-0.22	-0.34[e]
LPA	0.21	-0.12	-0.04
TPA	-0.07	-0.13	-0.22
9-Year-Olds (n = 50)			
MVPA	0.23	-0.30[e]	-0.49[e]
LPA	0.18	-0.05	-0.06
TPA	-0.07	-0.11	-0.30[e]
10-Year-Olds (n = 55)			
MVPA	0.12	-0.22	-0.37[e]
LPA	0.08	-0.18	-0.09
TPA	-0.11	0.20	-0.40[e]

[a]Sum 5 SF — sum of triceps, biceps, subscapular, abdominal, and medial calf skinfolds.
[b]MVPA — moderate to vigorous physical activity.
[c]LPA — low physical activity.
[d]TPA — total weekly physical activity.
[e]Statistically significant — p < 0.05.
Source: Compiled from Raudsepp, L., and Jürimäe, T., *Biol. Sport*, 13, 279, 1996, and from Raudsepp, L., and Jürimäe, T., *Am. J. Hum. Biol.*, 9, 513, 1997.

Table 3.7   Zero-Order Correlations Between Physical Activity and
Selected Anthropometric Parameters in Prepubertal Girls

Physical Activity	Stature	Body Mass	Sum 5 SF[a]
7-Year-Olds (n = 48)			
MVPA[b]	0.13	-0.32[e]	-0.49[e]
LPA[c]	-0.20	-0.16	-0.04
TPA[d]	-0.10	-0.18	-0.31[e]
8-Year-Olds (n = 54)			
MVPA	-0.07	-0.23	-0.54[e]
LPA	0.19	-0.12	-0.22
TPA	0.04	-0.21	-0.29[e]
9-Year-Olds (n = 57)			
MVPA	-0.05	-0.35[e]	-0.44[e]
LPA	0.12	-0.07	-0.19
TPA	0.12	-0.27	-0.30[e]
10-Year-Olds (n = 56)			
MVPA	0.18	-0.31[e]	-0.57[e]
LPA	-0.14	-0.06	-0.12
TPA	0.06	-0.40[e]	-0.51[e]

[a]Sum 5 SF — sum of triceps, biceps, subscapular, abdominal, and medial calf skinfolds.
[b]MVPA — moderate to vigorous physical activity.
[c]LPA — low physical activity.
[d]TPA — total weekly physical activity.
[e]Statistically significant — $p < 0.05$.
Source: Compiled from Raudsepp, L., and Jürimäe, T., *Biol. Sport,* 13, 299, 1996,
and from Raudsepp, L., and Jürimäe, T., *Am. J. Hum. Biol.,* 9, 513, 1997.

Physical activity increases energy expenditure and creates a negative energy balance, facilitating weight loss. Exercise increases the level of fitness and may affect many diseases associated with obesity. Researchers have conducted only limited controlled studies on physical activity in pediatric obesity. However, recently an excellent review article was published by Epstein and Goldfield.[187] The influence of physical activity on the level of body mass in children has been studied in three different ways:[187]

- Exercise influence compared with no exercise controls
- Exercise and diet influence compared with only dietary controls
- Different types of exercise programs compared

Results from the first type of study are contradictory. Gutin et al.[251] found significant reductions in body fat and increases in fitness levels compared with controls in 7- to 11-year-old children using aerobic exercises five times a week for 40 minutes per session. Conversely, Blomquist et al.[70] did not find any changes in 8- to 9-year-old children as a result of a training program. Results are also contradictory from the second type of study. Some studies indicate significantly greater changes in body composition for the diet and physical activity group in comparison with the diet-only group,[282] while other studies did not find significant differences between groups.[184] Epstein et al.[185] indicated that diet and physical activity combined are more effective than diet alone in increasing fitness levels. Only a few studies have compared different exercise programs for the reduction of body mass. Most exercise programs have focused on aerobic exercises. However, the optimal intensity and duration of these programs for children are not yet known. The best schedule for increasing intensity or duration of aerobic activity needs to be determined.

Few data address the use of resistance training in pediatric populations for increasing lean body mass and total energy expenditure. The best result may be achieved by combining aerobic and resistance exercises. In the U.S.,[251] data from several trials incorporating moderate to intense aerobic exercises suggest that school-based exercise intervention may provide a promising treatment for childhood obesity. Is it enough for children to exercise only moderately? The caloric costs of these kinds of exercises are relatively low (especially when the duration of the exercise is not long enough), and exercising may only increase the appetite. Physical activity patterns in children are characterized by short bursts of predominantly anaerobic activities. Large parts of children's physical activities are connected with different types of play. It is important to investigate the efficacy of school-based exercise programs since they provide an opportunity to develop healthy, active lifestyles in a large number of children.[638] The use of family to support activity programs may also prove useful for long-term change, since parental activity levels are strong predictors of child activity.[220] The development of physically active lifestyles has the potential for multiple benefits on obesity, comorbid physical and psychological problems, and acquisition of an active lifestyle that may accrue lifelong health benefits.

Table 3.8   Zero-Order Correlations between Physical Activity
and Eurofit Test Results in Prepubertal Boys

Variables	7-Year-Olds (n=53)		8-Year-Olds (n=45)		9-Year-Olds (n=50)		10-Year-Olds (n=55)	
	MVPA[a]	TPA[b]	MVPA	TPA	MVPA	TPA	MVPA	TPA
Standing long jump	0.22	0.13	0.18	-0.03	0.34[c]	0.22	0.29[c]	0.05
10 x 5 m shuttle run	-0.26	-0.18	-0.20	-0.04	-0.18	0.10	-0.06	-0.22
Bent arm hang	-0.18	0.04	0.22	-0.12	0.10	0.06	0.23	-0.02
Sit-and-reach	0.19	-0.17	0.22	0.04	0.12	0.01	0.20	-0.05
Plate tapping	0.04	-0.13	-0.06	0.23	-0.24	0.18	-0.10	0.19
Flamingo balance	-0.06	0.11	-0.02	0.18	0.13	-0.16	-0.12	0.20
Handgrip strength	0.14	-0.09	0.11	-0.18	0.18	-0.23	0.12	-0.03
Sit-ups	0.05	-0.12	-0.04	0.18	0.20	0.12	0.32[c]	0.15
20 m endurance shuttle run	0.22	0.18	0.30[c]	-0.09	0.40[c]	0.22	0.44[c]	0.30[c]

[a]MVPA — moderate to vigorous physical activity.
[b]TPA — total weekly physical activity.
[c]Statistically significant — p <0.05.

Source: Compiled from Raudsepp, L., and Jürimäe, T., *Biol. Sport,* 13, 299, 1996, and from Raudsepp, L., and Jürimäe, T., *Am. J. Hum. Biol.,* 9, 513, 1997.

Study results from our laboratory indicate that physical activity is constantly and significantly related to only one component of motor ability — aerobic fitness — in 7- to 10-year-old prepubertal boys and girls (Tables 3.8 and 3.9).[496-499] Laboratory results show reasonably constant correlations between moderate to vigorous physical activity and aerobic fitness across age. However, nonsignificant associations found between total weekly physical activity and motor ability clearly indicate that certain physical activity intensities are needed to influence motor ability in children.

Table 3.9   Zero-Order Correlations between Physical Activity and Eurofit Test Results in Prepubertal Girls

Variables	7-Year-Olds (n=53)		8-Year-Olds (n=45)		9-Year-Olds (n=50)		10-Year-Olds (n=55)	
	MVPA[a]	TPA[b]	MVPA	TPA	MVPA	TPA	MVPA	TPA
Standing long jump	-0.04	0.17	0.24	0.07	-0.10[c]	0.05	0.22	-0.03
10 x 5 m shuttle run	-0.23	-0.10	-0.28[c]	-0.07	-0.32c	-0.24	-0.27[c]	-0.14
Bent arm hang	0.19	-0.09	0.06	0.17	0.20	0.12	0.30[c]	0.22
Sit-and-reach	0.15	-0.05	0.16	-0.09	-0.06	0.17	-0.05	0.18
Plate tapping	0.07	-0.17	-0.18	-0.14	0.17	-0.03	-0.02	0.18
Flamingo balance	-0.17	0.11	-0.06	-0.22	0.21	-0.16	0.02	-0.11
Handgrip strength	0.19	-0.02	0.21	0.07	0.19	-0.13	-0.09	-0.19
Sit-ups	-0.04	0.26	0.08	-0.24	0.19	0.17	0.22	0.06
20 m endurance shuttle run	0.54[c]	0.37[c]	0.38[c]	-0.28[c]	0.36[c]	0.22	0.43[c]	0.10

[a]MVPA — moderate to vigorous physical activity.
[b]TPA — total weekly physical activity.
[c]Statistically significant — $p < 0.05$.

Source: Compiled from Raudsepp, L., and Jürimäe, T., *Biol. Sport,* 13, 299, 1996, and from Raudsepp, L., and Jürimäe, T., *Am. J. Hum. Biol.,* 9, 513, 1997.

Previous data concerning the association between aerobic fitness and physical activity are inconsistent.[442] Some studies have demonstrated a significant relationship between aerobic fitness and physical activity,[5,21] while others have not.[340,436] These studies varied on a number of dimensions — sample size and selection as well as measurement and assessment of physical activity and aerobic fitness — which may account for the differing results. The results from our laboratory are important from the perspective of health-related fitness since the intensity of physical activity is important to the fitness–activity relationship. Several other investigations have also emphasized the importance of moderate to vigorous physical activity.[563,588] Prepubertal children are considered quite active, but data are not available to support the contention that children are sufficiently active to account for their high levels of cardiorespiratory fitness.[589] To date, the optimal amount of physical activity for children is unknown. However, some daily moderate to vigorous physical activity is recommended.[266,588] The results of our investigations in prepubertal children clearly indicate the need for a stronger emphasis on moderate and vigorous physical activities in everyday sports and leisure situations.

Data on the associations between physical activity and other components of health-related fitness are less extensive. Our laboratory results have demonstrated generally low and variable correlations between physical activity and various physical fitness items in prepubertal boys and girls (Tables 3.8 and 3.9).[496-499] Significant relationships were found between moderate to vigorous physical activity and standing long jump in boys 9 and 10 years old as well as between moderate to vigorous physical activity and sit-ups in boys 10 years old. Moderate to vigorous physical activity was significantly related to 10×5 meter shuttle run test results in girls 8, 9, and 10 year old. Significant correlation was also found between moderate to vigorous physical activity and bent arm hang in 10-year-old girls. Thus, significant correlations were found between the moderate to vigorous physical activity scores and the results of these motor ability tests in which the body mass was moved or projected.

Associations between moderate to vigorous physical activity and several motor ability items found in our investigations reflect the role of environmental factors on motor ability. Despite the lack of studies on the relationships between physical activity and motor ability in prepubertal children, some investigations have reported low to moderate but significant correlations between physical activity and strength indices.[563] The relatively low association between physical activity and motor ability measures may be due to the parental report used in our investigations, which focused on activities classified by intensity and not by type of activity.

We also studied the influence of physical activity on 174 10- and 12-year-old rural children using the parameters of physical fitness measured according to the Eurofit test battery.[193,311] We used a 7-day physical activity recall, modified from Godin and Shephard.[235] Stepwise multiple regression analysis indicated that total physical activity predicted 45 to 49% of the variance in endurance shuttle run, standing broad jump, and bent arm hang in 10-year-old boys. In same-age girls, significant relationships were found between total physical activity and endurance shuttle run and 10×5 meter shuttle run (25 to 29% of common variance). Low physical activity accounted for 46% of the variance in endurance shuttle run and 25% of the variance in 10×5 meter shuttle run results in boys and girls, respectively. Vigorous physical activity accounted for 27% of the variance in 10×5 meter shuttle run in boys and 57% of the variance in endurance shuttle run in girls. In 12-year-old boys and girls, physical activity scores moderately but significantly (14 to 23% of the variance) influenced the results of endurance shuttle run and 10×5 meter shuttle run.[311]

The level of physical activity is negatively related to the amount of subcutaneous fat in prepubertal children. Physical activity is moderately associated with aerobic fitness. However, moderate to vigorous physical activity is an important predictor of those motor abilities that require body mass movement or projection. Indicators of physical activity are generally not significantly related to those motor ability tests that require muscular strength, balance, flexibility, and speed-of-limb movement in prepubertal boys and girls.

3.8   General considerations

Childhood physical activity is important as a protective health-related phenomenon. The accumulated evidence should prompt public health officials to advocate increased daily physical activity and, therefore, improved physical fitness levels. Cross-sectional investigations demonstrate a wide variation in physical activity levels among children. When these physical activity levels are maintained in rank order from childhood to adolescence, those children initially observed to be inactive and/or unfit, relative to their peers, would predictably become inactive and/or unfit adolescents. Although relatively little is known about how well physical activity tracks from childhood into adulthood, early measurement and intervention as a strategy are suggested to ensure healthy levels of physical activity and physical fitness in later years.

Modifying physical activity behavior may have both acute and chronic (into adulthood) influences on the health of children. The activity in children is spasmodic. It seems that children rarely or never reach steady state at any physical activity. The time after school and during weekends spent outside is strongly predictive of activity in prepubertal children. Physical activity is an essential prerequisite of health since humans are born to be active. The more active the children are, the fitter they are. The promotion of regular physical activity in children should be a priority of all health professionals. However, before a relationship between the level of daily physical activity and health can be identified, a valid method for the assessment of daily physical activity is needed. Despite the considerable research in the area of the development of appropriate physical activity assessment methods, there is no universally accepted method for the assessment of physical activity in children. Further efforts should be made to develop valid methods of physical activity assessment. Suggested methods for the assessment of physical activity in prepubertal children are presented in Appendix 3.

No extensive information exists on the relationship between the level of physical activity, anthropometric, and motor ability development in prepubertal years. Physical activity should increase and/or improve the morphological and functional characteristics of children over and above that which accompanies normal growth and development. However, several questions arise concerning the specificity and amount of physical activity that is needed to have a beneficial effect on different motor ability tasks during childhood. To date, the results of different investigations demonstrate that the amount of moderate to vigorous physical activity is a significant predictor of one motor ability component, aerobic fitness, in prepubertal years. However, the physical activity/motor ability relationship is not entirely clear since genetic, maturational, and environmental factors all contribute to the motor development in children. Further longitudinal studies should be conducted to assess the relationships among physical activity, different anthropometric variables and motor ability parameters in children during prepubertal years.

chapter four

Motor abilities of prepubertal children

4.1 Introduction

Several excellent reviews have been published about the ontogenetic development of motor function and motor abilities in childhood.[407,573] Inherent in any fitness program is the notion that fitness is for a lifetime. Usually, motor ability tests are used to motivate children to achieve higher levels of fitness and to include optimal levels of physical activity in their present and future lifestyles. Children can learn to get fit and to stay physically active throughout life. Fitness testing should be only one part of a total physical education program. The information provided by fitness test results is the basis for teaching children to plan their own exercise programs. There are clear relationships between good health and good test results. Accordingly, health-related fitness test items are favored. If awards are used, they should be based on the good health philosophy. Whitehead[673] noted that positive feedback on fitness tests influences motivation due to increased perceptions of physical competence. Corbin et al.[132] did not recommend the use of exclusive normative performance awards.

Many definitions of physical fitness have been presented during the last 30 to 40 years (for a review, see Pate[469]). Physical fitness has usually been viewed as a multifactorial trait related to the capacity of movement. Specialists have defined physical fitness as a set of attributes that people have or achieve that relates to the ability to perform physical activity.[112,689] Traditional definitions do not encompass the health outcomes of physical activity, and they should focus on the health-related aspects of fitness.[469]

Health-related physical fitness has been defined as a state characterized by an ability to perform daily activities with vigor and traits and capacities that are associated with low risk of premature development of hypokinetic diseases (those associated with physical inactivity).[469] However, these rules and definitions are related to adults. Good recommendations or definitions are not available for children.

Prepubertal children differ radically from adults in their physical growth and their cognitive, social, and psychological status. Children at ages 9 to 12 differ from preschool children. They better understand what they can do during the testing, and it is possible to motivate them to increase exercise with maximal effort. Fitness tests that require a long time and that are painful and uncomfortable are not acceptable. Prepubertal children need more knowledge and experience before testing than older children. Differences in testing performance between children and adults generally result from biomechanical rather than physiological factors.[529] Prepubertal children are not ready for overexertion without skill and motivational strategies on the part of the tester; so fitness tests that need a maximal exertion are not valid or reliable in young children. It is difficult to motivate children who are physically inactive, who do not participate in indoor or outdoor physical activities voluntarily, and who do not like physical exertion that demands increased breathing and/or sweating. Prepubertal children learn elementary skills of health-related exercises playing different games. It is important to increase childrens' motivation and interest in different physical activities. Parents play a key role in developing such interest.

Pate[469] recommended the use of three different concepts of motor abilities in children — motor performance, physical fitness, and health-related physical fitness (Table 4.1). Motor performance is the broadest of the three concepts of motor abilities and is defined as the ability to perform physical skills and rigorous physical activities including those involved in sports and athletics.

Table 4.1 Components of Motor Performance,
Physical Fitness, and Health-Related Physical Fitness

Motor Performance	Physical Fitness	Health-Related Physical Fitness
Anaerobic power	-	-
Speed	-	-
Muscular strength	Muscular strength	Muscular strength
Muscular endurance	Muscular endurance	Muscular endurance
Cardiorespiratory endurance	Cardiorespiratory endurance	Cardiorespiratory endurance
Flexibility	-	Flexibility
Agility	-	Body composition

Source: Modified from Pate, R. R., *Quest,* 40, 174, 1988.

Fitness tests that are suitable for use in a school environment and that provide valid and objective measures of fitness are simply not available. Fitness tests determine the obvious at best, only distinguishing the mature (or motivated) children from the immature (or unmotivated) children. Standard or norm tables confuse the issue of relative fitness because tables constructed on the basis of chronological age cannot be used to legitimately classify children at different levels of maturity. Some children view fitness testing unfavorably, and the tests largely contribute to negative attitudes toward physical education.[389] It is unrealistic to expect a significant increase in the level of fitness in children within the limited time allocated during school. Teachers must ask themselves why they are testing children's fitness — and if the answer is for classification purposes, they would be better employed seriously addressing the problem of children's sedentary lifestyles.[13] In the U.S., Dinubile[166] noted that all schools should establish fitness testing programs for children that are based on health-related physical fitness parameters rather than on athletic performance variables.

4.2   Health-related physical fitness

Physical fitness, participation in physical activity, fundamental motor skills, and body composition are important contributors to the development of a healthy lifestyle among children. Physically fit children live with a smaller risk of developing serious health problems.

Physical fitness has at least two aspects — health-related fitness and performance-related fitness. These have some degree of commonality, but important differences need to be recognized by children and teachers. Both aspects depend on genetic endowments (growth pattern) and are influenced by biocultural and biosocial components.[573] However, health-related fitness tests should measure factors that are directly concerned with the health and wellness of the individual.

Pate[471] emphasized that the term health-related physical fitness encompasses only three components for which a relationship to disease prevention or maintenance of good capacity for daily living are indicated. These are cardiorespiratory endurance, body composition, and neuromuscular fitness. Only a few data have demonstrated these relationships in children compared to adults. Fox and Biddle[214] included muscle strength, muscle endurance, flexibility, and posture as components of health-related physical fitness. Performance-related fitness tests include measures of explosive power, agility, coordination, and speed. Some fitness tests may involve a specific motor skill level, such as the throw.[167] Although health-related physical fitness is an attractive idea, relationships between measures of health and fitness need to be delineated among children. Certain aspects of fitness may not relate to health among children as they do among adults; or, because of differences in developmental physiology, other relationships may be documented.[31] There is only one test battery available with health-related fitness norms for 6- to 9-year-old boys and girls.[527] The American College of Sports Medicine[9] defined essential components of health-related fitness as cardiovascular fitness, muscular strength, muscular endurance, flexibility, and body composition. Health-related fitness components and recommended tests are presented in Table 4.2.

Table 4.2   Health-Related Physical Fitness Components and Measurement Procedures

Components	Laboratory Tests	Field Tests
Cardiorespiratory endurance	maximal aerobic power (VO_{2max}); submaximal cycle ergometer tests (PWC_{170})	distance runs (mile, 1.5 mile, 9 min, 12 min); step tests; graded shuttle run
Body composition	hydrostatic weight; deuterium oxide dilution; potassium counting; bioelectrical impedance	skinfold thickness; body mass indices; girth measures
Flexibility	goniometric measures; Leighton flexometer	sit-and-reach; stand-and-reach
Muscular strength	isometric dynamometer; isokinetic dynamometer; isoinertial one repetition maximum; cable densitometer	pull-ups; modified pull-ups; sit-ups
Muscular endurance	repetitions or time to fatigue at set percentage of maximum force	pull-ups; modified pull-ups; sit-ups

Source: Modified from Baranowski, T., et al., *Med. Sci. Sports Exercise,* 24, S 237, 1992.

4.3   *Biological maturation and motor ability*

In 1951 Seils[581] noted a significant positive relationship between skeletal maturation and motor performance in 6- to 9-year-old children. Significant relationships between maturation and motor performance have also been presented by other researchers.[2,50,51,403,406,575] Katzmarzyk et al.[321] demonstrated that the variation attributable to skeletal age, independent of chronological age, is a significant predictor of motor fitness in children 7 to 12 years of chronological age; and they emphasized that chronological age and skeletal maturity very rarely progress at the same rates. Relationships among chronological age, skeletal age, and body size confound their individual effects on performance. Effects of skeletal age are expressed mainly through body size, and skeletal age influences motor fitness more than it influences muscular strength.[321] Little et al.[371] indicated that (with the exception of flexibility) running speed, functional strength, explosive strength, static strength, upper body power, and aerobic power improve significantly with maturation in girls. More mature girls perform significantly better than less mature girls.[371] Age-specific correlations between skeletal age and PWC_{170} in girls generally increase with age and reach maximum at 11 to 13 years. The age trends in the correlations are less clear for PWC_{150}.[50]

Significant associations exist among static strength, explosive strength, and running speed when a wide age range is considered in prepubertal children. However, when age-specific correlations are calculated, only static strength is associated with skeletal age at all age levels.[391] Several authors[107,575] note that the strength of the relationship between skeletal age and motor performance capacities declines considerably when stature and body mass are partialled out in prepubertal children. We agree with Mafulli,[394] who suggested that performance standards should consider the biological ages of children more than their chronological ages.

Relationships between biological maturation and indicators of health- and motor-related fitness have not been intensively studied. Some investigations are based on studies of young athletes.[57,403] Few data are available about the relationships between maturation and performance in non-athletic groups. Beunen et al.[58] indicated that there are significant relationships between skeletal age and PWC_{130}, $PWC_{150,}$ and PWC_{170}, and the relationships are highest at the age of 11 (r = 0.53 to r = 0.64). Moderate correlations between submaximal performance capacity and skeletal age have also been presented for young athletes.[83] Beunen et al.[58] studied the relationships between different health-related and performance-related fitness test results and skeletal age in 6- to 16-year-old girls. The results of the bent arm hang, leg lifts and sit-ups correlated negatively with skeletal age; but the relationships were relatively low. Static strength was related to biological maturation from the performance-related components of fitness. Multiple regression analysis demonstrated that body size, skeletal, age, or chronological age were not the most important predictors of physical fitness in girls. The interactions of stature, body mass, skeletal age, and chronological age accounted for less than 10% of the variance in most fitness items.[58] In contrast, the same interaction explained the largest proportion of the variance in a variety of motor- and health-related fitness tests in slightly younger boys.[52] Thus, skeletal age does not have the same predictive value in the fitness of girls as in boys; it is a more important predictor of fitness in boys.[58]

Jones et al.[308] studied relationships between biological age determined by Tanner stages[618] and results of vertical jump, hand grip strength, and 20-meter shuttle run test in 10- to 16-year-old boys and girls. Stage of sexual maturity was significantly correlated with all physical fitness measurements (boys: r = 0.56 to r = 0.73; girls: r = 0.24 to r = 0.46). Analysis of covariance revealed that, when stature and body mass were taken into account, significant differences were evident between sexual maturity stages in boys but not in girls. This suggests that increases in body mass and stature are primarily responsible for the variation in physical performance of girls throughout maturation, whereas there are some qualitative differences in performance due to other factors in boys.

Estonian researchers[655] used a modified meta-analysis to establish the concordance vs. discrepancy in chronological age periods characterized by increased rates of annual improvement of running speed, muscle strength and power, and aerobic endurance. The material for analysis came from 31 original studies and 11 review articles. Available research data from the studies in former Eastern Europe and the former Soviet Union were also included in the meta-analysis. All material was analyzed by using annual changes or by plotting the mean values against chronological age. The acceleration in annual improvement rate was detected by the time of the greatest inclination in the curve. The percent of reports that indicated an increased rate of improvement of motor abilities at a certain age was called the consensus index (CI), and that index was used throughout the study. The age period studied ranged from 6 to 18 years. Since most reports covered only a part of that period, ages were divided into the following three segments:[655]

- Up to the age of 10 years (arbitrarily called preadolescence period)
- From 10 to 12–14 years (intermediate period)
- From 12–14 to 18 years of age (adolescence period)

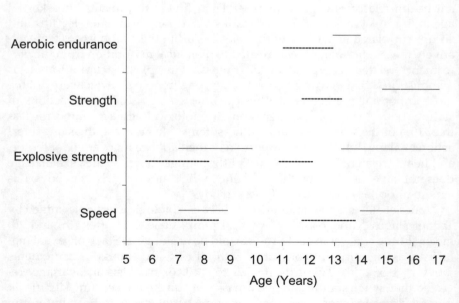

Figure 4.1 Periods of accelerated improvement of motor abilities in male (solid line) and female (dotted line) groups of children and adolescents. (Adopted from Viru, A., et al., *Biol. Sport*, 15, 211, 1998.) With permission.

Speed was expressed as a maximum running velocity (short dashes) from the motor tests[655] (Figure 4.1). The evaluation of power (explosive strength) resulted from standing long jump, vertical jump, or counter-movement jump. Muscle strength pattern resulted from handgrip strength, pulling strength of shoulders, arm pull, knee extension force, and clean and press. For the evaluation of aerobic endurance, only those tests were used that were of sufficient duration to assume the prevalence of oxidative metabolism. Accelerated speed improvement was found in boys aged 7 to 8 years and in girls aged 8 to 9 years (Figure 4.1). Accelerated rate of improvement in explosive strength was seen in boys 7 to 9 years old. In girls, satisfactory CI values were found at the age range of 6 to 12 years. No common period of acceleration of the improvement rate in muscle strength was found in preadolescents. The highest improvement in aerobic endurance occurred at the age ranges of 11 to 15 years and 11 to 13 years in boys and girls, respectively. Two accelerated improvements of motor abilities in children and adolescents were also found. In the male population, the periods from 7 to 9 and from 12 to 16 years were decisive for the improvement of motor abilities. In the female population, the first period of accelerated improvement occurred at the age of 6 to 8, while the second period appeared a year or two earlier than in boys. The authors of the meta-analysis emphasized that their results were in agreement with the results of most longitudinal studies regarding age characteristics of the increased rates of improvement in motor abilities.[655]

Several critical events take place that are essential for motor development during childhood and adolescence. In the period from 7 to 18 years, the most critical intervals seem to occur between 7 and 9 years and at the circumpubertal age. According to the theory of critical/sensitive periods in ontogenetic development, it is likely that the related critical events trigger the acceleration of improvement in motor abilities.[233,291,347,568,656]

The two critical periods require explanation. The intensity of the improvement of motor abilities declines at the ages closely related to the second and third stages of sexual maturation.[407] Sexual maturation may be associated with the phenomenon of "outgrowing one's strength," exhibited by a temporary inhibition or slowdown of motor development.[54] This phenomenon may not appear in all boys and girls.[54]

4.4 Anthropometry and motor ability

Anthropometrical parameters at the same chronological and biological age of children are very different. For example, the body stature in 10-year-old children could be different by more than 10 to 15 cm. The same is true with body mass. There are differences between boys and girls in prepubertal ages. Frequently physical education teachers note that, in the tests requiring speed and strength, the motor abilities in children with small statures test lower than children with normal or high statures. In contrast, the results of the tests that need endurance are relatively low in children who are very tall and have a high body mass.[168,586,595] The first data on the influence of anthropometrical parameters to different motor ability results were published more than 20 years ago by Malina,[397] Bouckaert et al.,[87] and Slaughter et al.[596]

Little information exists on the influence of stature and body mass (and especially the influence of skinfold, girth, length, and breadth/length parameters) to the different motor abilities in prepubertal children. In our laboratory, we have studied the relationships between anthropometrical profiles of boys and girls 10 to 12 years old (n = 70 and n = 69, respectively) measured by the protocols of the International Society for Advancement of Kinanthropometry[448] and Eurofit[193] tests results.[311] The following anthropometric parameters were measured:

- nine skinfolds — triceps, subscapular, biceps, iliac crest, supraspinale, abdominal, front thigh, medial calf, mid-axilla
- 13 girths — head, neck, arm relaxed, arm flexed and tensed, forearm, wrist, chest, waist, gluteal, thigh, thigh mid trochanter-tibiale-laterale, calf, ankle
- eight lengths — acromiale-radiale, radiale-stylion, midstylion-dactylion, iliospinale-box height, trochanterion-box height, trochanterion-tibiale laterale, tibiale laterale to floor, tibiale mediale-sphyrion tibiale
- eight breadths/lengths — biacromial, biiliocristal, foot length, sitting height, transverse chest, A-P chest depth, humerus, femur

Somatotypes were calculated according to the Heath-Carter method.[273] The Pearson correlation analysis indicated that Eurofit tests results did not correlate significantly with stature in boys. However, the results in standing broad jump (r = 0.27), sit-ups (r = 0.38), hand grip (r = 0.31), and 10×5 meter shuttle run (r = -0.24) were dependent on stature in girls. Stepwise multiple regression analysis indicated that stature influenced the results of the tests by only 5.80 to 14.16%. Surprisingly, body mass and BMI did not influence the results of Eurofit tests in girls; and there were only significant relationships between body mass and sit-ups (r = 0.27) and hand grip (r = 0.25) in boys. Finally, the relationships between stature, mass, and BMI were not better with Eurofit test results when we used the extreme stature and body mass in 25% and 75% groups of the children. Only two skinfold thicknesses characterized the endurance shuttle run test results at a relatively low level — 11.27% and 25.95% ($R^2 \times 100$) in boys and girls, respectively. Skinfold thicknesses characterized the results of the standing broad jump at relatively high levels in boys (27.51%) and girls (18.17%). The influence of skinfold thicknesses was relatively high on the tests results of hand grip (20.63%), sit-ups (23.78%) and Flamingo balance (22.95%) in girls. Girth parameters characterized 23.26% and 19.03% of the total variance of the Flamingo balance test results in boys and girls, respectively. Surprisingly, the girth values highly influenced (30.35% and 23.71%) hand grip strength and sit-ups results in girls. Different length parameters influenced more motor ability tests in girls in comparison with boys. For example, sit-ups (49.60%), Flamingo balance (30.83%), standing broad jump (23.81%), and hand grip (22.13%) results highly influenced different length parameters (mostly on the trochanterion length) in prepubertal girls. Only the results of the sit-ups test highly influenced (31.34%) breadth/length parameters in girls. Somatotype components did not influence the results of the Eurofit tests in boys (except ecto- and mesomorphy influenced the sit-ups test results by 13.15%). However, ecto- and mesomorphy influenced the results of the five Eurofit tests (20-meter endurance shuttle run, hand grip strength, standing broad jump, sit-ups and plate tapping) by 6.19 to 20.07%. Thus, relative linearity (ectomorphy) and robustness (mesomorphy) are the components that influence motor ability of prepubertal children, especially in girls. Endomorphy, which characterizes the body fat, did not influence different motor ability tests in prepubertal children (except plate tapping in girls, which was rather surprising). We concluded that the influence of anthropometric parameters to the motor abilities in prepubertal children is moderate. Our children had normal body statures and masses.[311]

4.5 *Physical activity and motor ability*

In children, increased physical activity and physical fitness are associated with improved health indices.[104,504,529] It is well known that health status is significantly correlated with physical activity.[85,261,480] However, the strength of this association is moderate at best in children.[322,404] Dennison et al.[153] indicated that physically active adults had significantly better childhood physical fitness test scores than did inactive adults. Levels of habitual physical activity vary during growth and aging, and specific measures of physical fitness vary with growth, maturation, and aging — independent of physical activity.[407] On the other hand, regular physical activity and lifestyle influence physical fitness from childhood through adulthood.[85,402,404,504]

There are as many types of health-related physical activities as there are types of physical fitness related to health. Some of the documented and suspected relationships among physical activity, physical fitness, and health are presented in Figure 4.2.[31] Recently, researchers concluded that it is more important to monitor childrens' participation in physical activity than to monitor fitness for public health purposes.[689] In their review article about the relationships between motor ability and aerobic fitness in children, youth, and adolescents, Morrow and Freedson[442] concluded that there is a small to moderate relationship between these parameters. They suggested that the weak association identified may be due to poor measurement of physical activity, high level of aerobic fitness, and/or the lack of a relationship in the first place.[442]

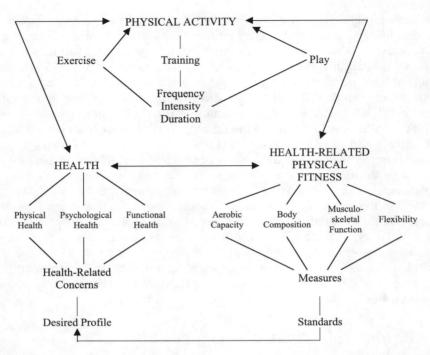

Figure 4.2 Relationships among health, physical activity and health-related physical fitness.

Source: Modified from Bouchard, C., and Shephard, R. J., Physical activity, fitness, and health: the model and key concepts, in *Physical Activity, Fitness, and Health,* Int. Proceedings and Consensus Statement, Shephard, R.J., and Stephens, T., Eds., Human Kinetics, Champaign, 1994, 77.

Armstrong et al.[15] found no relationship between habitual physical activity as estimated from continuous heart rate monitoring and directly measured oxygen consumption in 11- to 16-year-old British children; while Rowlands et al.[534] indicated that there is a positive relationship between physical activity and aerobic fitness and a negative relationship between fat and physical activity in 8- to 10-year-old children. In another study, poorer fitness was found in 9-year-old children who were shorter, obese, and less active than their counterparts.[49] Fenster et al.[203] suggested that there are great difficulties in correctly measuring aerobic capacity and physical activity in children. Their study demonstrated a significant relationship only between aerobic capacity measured by VO_{2max} and the level of physical activity measured by large-scale integrated activity monitors in 6- to 8-year-old children.[203]

Some studies divided physical activity into subclasses by intensity that correlated with VO_{2max} in children.[5,21,252,473,532,538,613] Daily physical activity of moderate to high intensity has been shown to correlate significantly with

VO_{2max} in prepubertal boys.[21,473,613] However, Al-Hazzaa and Sulaiman[5] indicated that VO_{2max} demonstrated a significant relationship with the percentage of time spent at activity levels above the heart rate of 169 beats per minute — but not with activity levels above the heart rate of 159 beats per minute — in 7- to 12-year-old Arabic boys.

Pate et al.[473] studied associations between two measures of physical fitness — 1.6 km run/walk performance and sum of three skinfold thicknesses — and selected physical activity factors in an American representative sample of third- and fourth-grade children (8 to 9 years old, with a total of 1150 boys and 1202 girls). Physical activity variables were used in two multiple-regression analyses in which run/walk time and sum of skinfolds were dependent variables. Multiple R^2 for these two analyses were 0.21 and 0.18. The results indicated that physical activity and physical fitness were significantly, although moderately, associated in young children. They concluded that interrelations directed toward enhancement of physical activity in children are worthy of investigation.[473]

Very few have studied the relationships between physical activity and motor abilities except aerobic fitness. Sallis et al.[550] examined the relationship between habitual physical activity and components of health-related fitness (one-mile run, pull-ups, sit-ups, sit-and-reach) in fourth-grade children. The physical activity index was significantly associated with the results of sit-up and sit-and-reach tests in boys and girls as well as the pull-up test in boys. Thus, childhood physical activity is associated with all fitness components with some gender differences — canonical correlations indicated that the association is slightly higher in boys than in girls.[550]

Dennison et al.[153] hypothesized that childhood tests of physical fitness predict adult physical activity levels and that other variables are also important determinants of adult physical activity levels. The Youth Fitness Tests results at the age of 10 to 11 years were compared with the level of physical activity at the age of 23 to 25 years. The results demonstrated that physical fitness testing in boys helped to identify those more likely to become physically inactive young adults. Fitness tests that measured endurance (548.6 meter run and maximum number of sit-ups) had the strongest relationship to risk of adult physical inactivity. Tucker and Hager[639] studied the relationships between watching television and muscular fitness (pull-ups, push-ups, and sit-ups) in 9- and 10-year-old boys and girls. They found no significant relationships between time spent watching television and muscular fitness test results.[639]

Several studies indicate that aerobic fitness is moderately correlated with physical activity in children. However, there is a lack of data about the relationships between physical activity and other health-related physical fitness parameters in children.

4.6   Genetic determination of motor ability

Health-related and performance-related fitness have not been widely studied in the genetic context. Many have studied the heritability of aerobic fitness, primarily using the measurement of VO_{2max}.[82,183,391] Eystron and Fischbein[183] and Bouchard[82] indicated a moderate genetic effect of about $r < 0.40$ on VO_{2max}. In 10-year-old twins, Maes et al.[391] found slightly higher heritability coefficients. Few studies have investigated the heritability of other health-related fitness variables.[345,391,477] Kovar[345] and Maes et al.[391] found that flexibility was highly heritable. Conflicting results are published about the heritability of the sit-up test. The heritability coefficient $r = 0.69$ was higher in Kovar[345] study compared with Perusse[477] results ($r = 0.21$). The bent arm hang test heritability coefficients are reportedly between $r = 0.35$[345] and $r = 0.65$.[391]

Vertical jump tests have a high genetic component in performance-related fitness variables.[345] Results are similar for the shuttle run test, with coefficients varying from $r = 0.72$ to $r = 0.90$.[345] The mean heritability estimate for static strength measured with different strength tests is somewhat lower.[183,345,477] However, Maes et al.[391] suggested a higher heritability for static strength and running speed ($r = 0.71$) than for explosive strength in 10-year-old children. The heritability coefficient of $r = 0.51$ is presented for balance.[391]

The Maes et al.[392] study focused on the qualification of genetic and environmental sources of variation in physical fitness components in 105 10-year-old twin pairs and their parents. Motor ability tests were divided into performance-related and health-related tests. Performance-related fitness characteristics were moderately to highly heritable. The heritability estimates were slightly higher for health-related fitness characteristics.[392] Having studied about 600 literary sources on human genetic development, the Ukraine researcher Sergienko[572] concluded that human morphological characteristics are more herited than different motor ability parameters.

4.7   Competitive sports and motor ability

It is increasingly clear that peak athletic performance can only be obtained if systematic training begins early in life. The potential benefits of organized sports for children include: (1) improvement of health; (2) enhancement of normal physical and social growth; and (3) enhancement of maturation. Organized sports also improve motor skills and physical fitness in children, particularly in those who are physically and mentally challenged. Organized sports competitions in children can, if properly structured, play an important role in socialization, self-esteem, self-perception, and improved psychological well-being. Organized sports can also establish the basis for a healthy lifestyle and lifelong commitment to physical activity.[603] Much controversy still exists regarding competitive sports participation at a young age, including ethical and moral issues of early specialization in one sport.[503]

Some sport events — swimming, gymnastics, tennis — require that children begin intensive exercise during the prepubertal ages. Children sometimes begin their sport at the age of 4 or 5 years. However, sports specialization should be avoided before the age of 10. Are children who begin systematic exercise 1 to 3 years earlier more fit than children who have not yet begun a sport? In Australia, several groups of prepubertal swimmers and tennis players were studied under the University of Western Australia Growth and Developmental Study program about 20 years ago.[69] Differences in motor ability test results were not high between children engaged in sports and children not active in sports.

Two excellent review articles have been written by Malina[399,400] about the problems of youth sports. It is not easy to determine whether a child is ready for sport. What are the criteria? Several components are important — physical, motor, social, emotional, and cognitive. The ability of a child is biocultural, and no single answer exists about readiness. Readiness is a functional concept that emphasizes the relationship between the ability of an individual and the demands of a specific activity or task. Readiness is related to the theory of critical periods — specific times during which the child is maximally sensitive to enviromental influences, both positive and negative, during growth and maturation and during the development of skills and behaviors.[399,400] Critical periods may represent times of maximal readiness.

Talent selection for top sports was highly organized in the former Eastern European countries (especially in the former Soviet Union and the German Democratic Republic). Special selection criteria for gymnastics were presented;[262] for example, children began their careers at about 6 to 7 years of age. A primary selection phase for talent in Rumania targeted children between 3 and 8 years of age, while the more important secondary phase varied according to the sport — 9 to 10 years for gymnastics, figure skating and swimming, and, for other sports, 10 to 15 years for girls and 10 to 17 years for boys.[77] Baxter-Jones[44] suggested that more mature children could participate in sports needing power and speed, while less mature children could participate in sports such as dance and gymnastics.

One of the most important components of the selection criteria is the measurement of different motor abilities. For example, the Canadian Talent Identification Program for female gymnasts[25] consists of glide kips, 20-meter run, leg lifts, standing long jump, chin-ups, vertical jump, push-ups, hip pullover, and rope climb. The best way to select talent is to use the testing results from schools and add event-specific tests.

Reilly and Stratton[503] argued that there are few, if any, models of talent identification and nurturing that are globally acceptable. Du Randt[171] also noted that talent identification is uncoordinated and under-researched — although there is a definite need for it, especially for identification in accordance with scientific methods. Du Randt[171] suggested that the first stage of identification should take place at the age of 8 to 10 in the form of mass screening (this age can vary depending on the sport), and this should be followed up

18 to 24 months later. Final talent identification should take place at around 14 years. However, Arnot and Gaines[20] stated that the talent should be recognized and encouraged in children after the age of 10, since such talent is an important part of the overall potential of a child and one that deserves recognition and encouragement as much as any other. The degree of maturation has been reported to highly influence talent selection.[503] For example, Reilly and Stratton[503] indicated that early-maturing males are at an advantage in many sports because of their significant increase in muscle mass during peak growth.

Sports activity may influence pubertal development, sexual maturation, and its major event in girls, menarche.[410,612] Biological maturation in girls is delayed, for example, in gymnasts.[101] Investigations have indicated a delay in skeletal maturation of 1.3 years in female rhythmic gymnasts.[229] This is likely due to the late biological maturation and a combination of very intensive training and varying biological and social factors.

The results of the body fat content in young soccer players are controversial. Data indicate that young soccer players have more body fat compared with norms,[659] or there are no differences in body fat content between prepubertal soccer players and a control group. Young (10 to 12 years old) elite soccer players are leaner than the same age non-elite players.[257]

Children begin to exercise at the age of 4 to 5 years in some sports events. Selected children begin to exercise with a high training load, which increases the possibility of overtraining and/or injuries. However, motor test batteries for sport talent selection are not yet available.

4.8   Measurements methods

4.8.1   Main criteria for motor ability tests

Physical fitness is one facet of sports and physical activity that can have short- and long-term influence on health and well-being in children. Fitness tests have been used in a variety of contexts to plan curricula in schools, direct social and government health care practices, and to determine readiness for combat.[539] One consideration in fitness testing is that poor selection of test items may produce harmful consequences, especially items involving twisting of the spine.[366]

Several motor ability tests have been presented by physical education teachers, pediatricians, exercise physiologists, sport physicians, and coaches during the last 100 years. Little in the field of physical education has stimulated as much emotional debate as components, interpretation, and value of physical fitness testing. For example, Pate[470] stepped out of his physiologist role to advocate the pedagogical uses of motor ability tests, while Seefeldt and Vogel[571] took a strict measurement position by suggesting that fitness tests are psychomotorically unsuitable for use in children. In contrast,

Whitehead et al.[675] indicated that, while there is certainly justification for halting the isolated and inappropriate use of fitness tests, there were also good reasons for advocating the use of field tests of health-related physical fitness as curricular tools within comprehensive fitness education programs. Some researchers have indicated that motor ability tests are not educational objectives but rather are tools that may be used in the curricular process of attaining them.[214,215] However, when motor ability tests can be used to cultivate each individual's sense of physical self-worth, then they also have an enhanced positive effect.[8,60,216] One of the criteria in selecting test items is to find items that measure differences of physical fitness. Each item in the test adds new information about the children. Finally, tests must be reasonably familiar to physical education teachers, economical in terms of time and expense, and feasible to administer in field situations.

Very few investigations on motivational outcomes of motor ability testing and award schemes have been completed. However, there have been a few attempts to apply the tenets of well-tested theories from mainstream psychology to youth fitness testing.[8,60,216] Self-evaluations of personal competence invalues interpersonal comparison of abilities while others focus on task mastery.[172,447] An example of interpersonal ability comparison in motor ability testing would be the use of normative tables for test score interpretation, while an example of task mastery evaluation would be comparing pre- and post-tests of fitness after the exercise program. However, a low ranking on tests might have a negative effect on intrinsic motivation.[648] Positive emotions in testing are very important in young children. Task mastery maximizes the likelihood of improving self-competence perceptions since the maintenance and/or improvement of fitness is inevitable when children exercise regularly, especially children who initially have low fitness levels, because they often respond well to training. Thus, the focus on the exercise process rather than the fitness product has been recommended in education programs.[131]

In the U.S., children have been offered awards for scoring at or above particular percentile levels of motor ability tests. These awards may be perceived as an extrinsic or controlling reason for training for and doing the test, and this may reduce intrinsic motivation.[132,215] Also, a sense of personal success is minimized because so few children can win the awards.

Motor ability tests recommended for use in schools are aimed at measuring abilities such as endurance, strength, flexibility, etc., rather than measuring skills. However, field motor ability test results do not frequently correlate significantly with laboratory tests. For example, strength measured by means of a field test in an isoinertial context does not reflect isometric or isokinetic force measured in the laboratory.[329] Most laboratory tests are too cumbersome for use in field conditions.[214]

Bovendeerdt[88] presented three recommendations for physical education teachers before testing children:

- Physical fitness tests can only be used by teachers of physical education who believe that fitness is an explicit goal of their teaching.
- Teachers of physical education must learn how to accurately measure and organize their testing effectively in school classes.
- Before testing, teachers of physical education must consider how to use the results they get in their testing activities.

A motor ability test result is acceptable only when the child tries to do his or her best. Otherwise the reliability of the test is a problem. Maximal exertion could be the main reason why some tests have insufficient reproducibility (especially most endurance and strength tests). An acceptable warm-up, explanation, and demonstration before testing are also needed to prevent injuries.

The measurement of physical fitness raises several conceptual, methodical, and technical problems that explain why surveys including such measures have been scarce until recently. The main problems linked with fitness measurements are validity, reproducibility, reliability, normalization, and standardization. The three most important characteristics of the tests are validity, reliability, and objectivity. Validity refers to the degree to which a test measures what it is supposed to measure and is the most important characteristic of testing.[39,142,258,297] In field testing, the biggest problem is the objective measurement tool for establishing construct validity of the criterion variable, which is the score used to represent a participant's ability at some part of a whole skill or an abstract trait. Reliability refers to the degree of consistency with which a test measures what it measures — essentially, repeatability of the test.[39] When children are measured two times with a perfectly reliable test, the two scores will be nearly identical. Objectivity is the degree to which multiple scores agree. This is also known as interrater reliability or interrater agreement.[39] To enhance test objectivity, it is essential to have a clearly defined scoring system and a competent scorer. The scorer should be trained and experienced with the testing instruments used.

Reliability and validity are the main challenges when recommending test batteries to schools. Specific tests could be appropriate for one purpose but inappropriate for another.[675] Compared with laboratory tests, it is probably true that recommended field tests for children do not have sufficient validity and/or reliability. However, research data[6,535] suggest that, in most cases, health-related physical fitness tests results have enough reliability and validity to:

- Enable reasonable diagnosis of fitness needs
- Facilitate development of appropriate exercise prescriptions
- Establish baselines for future evaluation of fitness goal attainment

Validation of field tests remains difficult.[221,329] For example, validation of such tests as bent arm hang, leg lifts, arm pull, and standing long jump demonstrates that strength and muscular endurance factors measured by these field tests cannot be compared with isometric strength testing in laboratory conditions in 12- to 18-year-old boys and girls.[329]

Corbin and Pangrazi[134] claimed that poorly constructed fitness standard surveys of youth fitness may account for low scores by children on specific fitness tests. Updyke[645] believed fitness tests served to motivate individuals. However, fitness tests can deter motivation when standards are high. Questions remain as to how to individualize criteria for children. Blair[64] stated that, when 20% of children are at risk because of low fitness, then major efforts should be made to correct deficiencies rather than making general statements about low youth fitness.

Different measures exist for different types of fitness, and they often have clear criterion measures and useful field measures for each type.[221,296,330,471,535,598] However, the selection of a test from different batteries is very complicated (see Chapter 4.8.2). Relatively easy, non-time-consuming, valid, and reliable test batteries are needed for large-scale studies.

Once test items have been selected to measure components and subcomponents of physical fitness, standards are selected to serve as bases for evaluating fitness from a health status standpoint. Standards are expressed as general or specific. A general standard is associated with the general population. A specific standard is adjusted in some way to account for effects of a specific impairment upon performance. Specific standards are only provided for selected test items for specific target populations. If the type of standard is general, physical educators typically have two levels of standards from which to choose — minimal and preferred. A minimal standard is considered an acceptable score that meets the lowest acceptable criterion of health. Most children should be able to attain the appropriate minimal standard. A preferred standard conveys a higher level of fitness and is, therefore, more desirable. A preferred standard represents a good level of fitness and is one most children find challenging. As an example, in the well-known Fitnessgram test battery, both minimal and preferred standards are presented for some tests.[129] If the standards are unavailable or inappropriate for specific children, teachers are encouraged to develop individualized standards by which to assess performance. An individualized standard is a desired level of attainment for an individual in an area of health status that is established in consideration of the present level of performance and progress toward a specific or general health-related standard.

The so-called norm-referenced standards and criterion-referenced standards are today used to interpret individual fitness test results.[221] Norm-referenced standards use age and gender-specific population distributions to judge fitness tests results. A nationally sampled reference group is generally used as the standard, and an individual's score is compared to the norms. The test score for an individual is expressed as a percentile. Children who perform at the 50th percentile are usually rated as having achieved an adequate level of fitness. Currently, this kind of evaluation is not widely recommended because it does not consider health-related factors. It is probably incorrect to use national sample reference groups for presenting standards, especially in big countries. Large differences may exist between regions on racial, socioeconomic, and climatic conditions in big countries. Criterion-referenced standards set specific cut-points or ranges for acceptable test scores, or cut-point or range values that are believed to be associated with current health and the ability to carry out functions of daily living.[143] Three limitations of the criterion-referenced standards approach include:[143]

- The setting of a standard is somewhat subjective.
- Youths may be misclassified.
- Each standard does not offer adequate incentive for maximal achievement.

Normative data alone do not allow qualitative interpretation of the fitness levels of children; and criterion-referenced standards that define levels of fitness associated with quantifiable public health benefits are needed.[274,588] Some standards have been proposed in the U.S.,[11,143,292-294] but no consensus has been reached.[540] Optimal standards await the results of controlled longitudinal studies that measure the strength of association of levels of health-related fitness among youths both to adult risk factors in youth and to levels of cardiovascular disease morbidity and mortality when youths become adults.[31]

It is difficult to characterize different test results in children. The test score might result from a combination of the children's ages, maturity levels, genetic backgrounds, skills at the specific test, and preliminary training.[84,120,214,216,346] Most tests (or test batteries) have some form of age- and sex-dependent normative standards by which children can compare their performances to their peers. Norms are usually expressed as percentiles or standard scores. When performance is converted to a normative standard, it is referred to as norm-referenced measurement. Admittedly, it is almost impossible to set definitive criterion standards for interpretation from the strict measurement perspective. However, from our educational perspective, doing so is preferable to the alternative methods — by normative comparison, through individual teacher judgments, or with no interpretation at all. Current criterion-referenced standards established for children are less than perfect and will no doubt need adjustment when data from longitudinal studies become available. In addition, criterion-referenced standards have less potential for producing negative psychological outcomes than the interpersonal comparisons inherent in normative ranking.[133,215] However, several physical education specialists believe that peer comparison is useful for discriminating within and among ability groups.[540] On the other hand, peer comparison is inappropriate for interpretation of performance on health-related fitness or motor fitness performance tests because the results of such tests largely depend on maturity and genetic endowment rather than a commitment to regular physical activity by the child.[214-216,571] Children may perceive norms beyond their capabilities or competence and consequently withdraw from exercise.[132] Children with the greatest need — those with low fitness levels — need to be identified so that special programming can be developed for them.[65] National reference standards are not recommended for these children, and the calculation of changes compared with previous testing is suggested. It is positive to present changes during one year or 6 months. These methods can also be used for testing late maturers.

When comparing elementary and secondary schools, the health-related fitness testing in elementary schools is more complicated. Testing is not adopted on a broad basis in many schools, and the value of their physical education programs cannot be assessed.[331,415,625] Few physical education teachers like giving fitness tests or have the preliminary training to administer the tests. For programs to succeed, the amount of time, the number of qualified instructors, and the available facilities must be improved; and most of these factors are not under the control of physical education teachers.[625]

4.8.2 Motor ability test batteries

A good field test does not require expensive equipment that can only be used by highly trained personnel in controlled settings. Most field tests of health-related physical fitness can be used in a variety of settings, including schools. Evidence must show that the recommended field test is a valid measure of some aspect of physical fitness — perhaps not as valid as the laboratory test, but with acceptable validity nonetheless. The same tests can be administered to both boys and girls. An excellent historical review article about youth physical fitness awards in the U.S. was presented more than 10 years ago by Corbin et al.[132] These awards were established to motivate improvements in youth fitness, to motivate youth to want to take fitness tests, and to encourage active lifestyles. Whitehead et al.[675] suggested that physical education teachers today are confused about which tests should be used and why. This confusion is fueled by philosophical debate and criticism of the tests from a measurement perspective. A clear need exists for a radical reappraisal of what contributes to appropriate or inappropriate use of fitness tests.

It is not easy to compare the motor ability test results in children. The most important questions when comparing obtained data are the following:

- Are the test batteries (and items) valid, reliable, and standardized in all compared groups (different countries)?
- Are there differences in race, gender, age, or school grade?
- What are the aims and purposes of the different studies?
- How old are the measurement results compared?
- Are the results longitudinal or cross-sectional?
- What is the physical activity level of the children whose motor abilities are compared?
- Are the children measured in the same season (winter or summer, hot or cold)?
- Are there any differences in chronological and/or biological ages?

Motor abilities of American and European children were compared for the first time in the 1950s.[349] The final result of this comparison was that children in America were relatively unfit in comparison with European children. The national norms for Americans who took the AAHPERD (American Alliance for Health, Physical Education, Recreation and Dance) Youth Fitness Test were presented in 1958.[571] The Fitnessgram test battery was included in the mid-1980s to help teachers to improve the fitness of children.[294] AAHPERD then appointed a new task force and writing team who published the Physical Best package in 1988,[7] which was very similar to Fitnessgram tests. The President's Council on Physical Fitness and Sports presented its test battery in 1987 as a separate entity.[483] The tests used in different recommended test batteries are presented in Table 4.3.[541]

Table 4.3   Motor Ability Test Batteries Recommended in the U.S.

Item	Test battery					
Fitness Tests	Fitnessgram	President's Challenge	YMCA Youth Fitness	National Youth Fitness Program	Chrysler AAU	Physical Best
Walk/run	X	X	X		X	X
Shuttle runs	X	X	X		X	
Skinfolds	X		X			X
Sit-and-reach	X	X	X		X	X
V-sit		X				
Sit-ups				X	X	X
Curl-ups	X	X	X	X		
Pull-ups	X	X	X	X	X	X
Flexed arm hang	X			X	X	X
Push-ups				X	X	X
Standing long jump						X
Others	Trunk lift; Hip flexor test; Body mass index				Hoosier endurance run; Phantom chair; Sprints	Body mass index

Source: Modified from Safrit, M. J., *Complete Guide to Youth Fitness Testing,* Human Kinetics, Champaign, 1995, 42.

All fitness test batteries recommended in the U.S. are more alike with regard to the components of physical fitness they measure. All test batteries include tests for aerobic capacity, flexibility, and abdominal muscular strength and endurance (Table 4.3). Upper body strength is measured in five test batteries, and agility is measured in two test batteries (but is recommended only for young children in one test battery). All test batteries include distance run tests for one mile or longer, with the exception of shorter distances for young children in two instances. Some differences also exist in flexibility tests. The well-known sit-and-reach test has most frequently been used (in some cases in inches and in some cases in centimeters). The V-sit is recommended in one test battery. Retest variability likely occurs in body strength measures. Variations include regular pull-ups, a pull-ups/flexed arm hang option, and a modified pull-ups test. Body composition is identified as an important component of fitness in four of the six tests. The triceps and calf sites are measured in the sum of skinfold tests in all batteries dealing with body composition. However, BMI is offered as an alternative to the skinfold measures in two instances. The President's Challenge Physical Fitness is a U.S. award — a popular, long-standing recognition program for school-aged children. The award is offered to young people who undertake a battery of physical fitness tests. For example, approximately 10 million students received one of the four awards (presidential, national, participant, or health fitness) during the 1998–1999 school year.[602]

The relatively new Brockport Physical Fitness Test (BPFT) battery[682] can be used with the general population and with youngsters with disabilities aged 10 to 17. The BPFT includes a number of unique features. First, in an effort to personalize testing and assessment, the test battery includes 27 different test items (Table 4.4). However, a complete test battery for one individual or category of disability generally includes four to six items. Second, it applies a health-related, criterion-referenced fitness approach to youngsters with disabilities. Third, it provides an approach based on health-related needs and a desired fitness profile. Finally, many of the test items are new (or at least nontraditional) and include a larger number of youngsters with disabilities in a physical fitness testing program than previously reported.

Table 4.4 Components and Test Items
of the Brockport Physical Fitness Test

Body Composition
Skinfold measures
Body mass index

Aerobic Functioning
PACER test (20 m)
PACER test (modified 16 m)
Target Aerobic Movement Test
One-mile run/walk

Musculoskeletal Functioning
Muscular Strength/Endurance
Trunk lift
Dominant grip strength
Bench press
Isometric push-up
Push-up
Seated push-up
Dumbbell press
Reverse curl
Push/walk (40 m)
Wheelchair ramp test
Curl-up
Curl-up (modified)
Extended arm hang
Flexed arm hang
Pull-up
Pull-up (modified)

Flexibility
Back saver sit-and-reach
Shoulder stretch
Apley test (modified)
Thomas test (modified)
Target stretch test

Source: Modified from Winnick, J. R., and Short, F. X.,
The Brockport Physical Fitness Test Manual, Human
Kinetics, Champaign, 1999.

Canadians have been tested using the Canadian Association for Health, Physical Education and Recreation (CAHPER) Test Battery.[127,270] A total of 11,000 students aged 7 to 17 were measured in 1965.[270] The second survey in Canada was undertaken in 1981.[127] The *Manitoba Physical Fitness Performance Test Manual and Fitness Objectives* was published in 1977 and contained the results of a physical fitness survey of the Manitoba schools conducted from 1976 to 1977.[411] The Manitoba test preceded the CAHPER II test with the inclusion of a more valid measure of aerobic fitness as well as tests to estimate adiposity and flexibility. The Canadian fitness survey[106] has been presented for Canadians 7 to 69 years of age. Although the Canadian fitness survey included norms for younger children, it is currently recommended only for persons 15 years and older. The Canadian Home Fitness Test was developed in 1976 for the health and welfare of Canadians as a self-administered motivational test. One part of the test has been extended to the junior high school age range of 11 to 14.[23] An excellent monograph about the physical fitness of Canadians between the ages of 10 and 69 has been presented by Shephard.[574]

In Europe, the development of motor ability tests followed America with a delay of 20 years. Since the end of the 1970s, physical fitness tests with school children have been carried out in the Netherlands.[325,365,424] Kemper[325] developed a test battery called MOPER (MOtor PERformance) in 1977, which has been used since 1986 with a great number of schoolchildren. Quintile reference scales for the MOPER fitness tests have been used in the Netherlands. In Germany, Schneider[565] examined 10- to 16-year-old children with four fitness test items of the AAHPERD (1980) test battery. In the former German Democratic Republic, a Representative Test Battery was used in schools.[243] A large sample of school children (n = 3000) aged 7 to 16 was tested in a mixed longitudinal and cross-sectional study between 1968 and 1986. Sport-related tests were mostly used.[243]

Using different test batteries, it is very clear that a comparison of data fails due to the differences in fitness test items used in different investigations. Most of the tests were also performed in different ways. Some problems occurred with translations and understanding that led to a different administration of a test item. This problem became more significant when studies did not employ pure fitness tests and added questionnaires. Manuals

Table 4.5   Dimensions and Factors of Physical Fitness of the Eurofit Tests[a]

Sequence of Testing	Eurofit Test	Dimensions	Factor
1	skinfold thickness; biceps; triceps; subscapular; calf	body composition	body composition
2	flamingo balance	balance	total body balance
3	plate tapping	speed	speed of limb movement[a]
4	sit-and-reach	flexibility	flexibility
5	standing broad jump	strength	explosive strength (power)[a]
6	hand grip	strength	static strength[a]
7	sit-ups	muscular endurance	trunk strength
8	bent arm hang	muscular endurance	functional strength[a]
9	shuttle run 10 x 5 m	speed	running speed; agility[a]
10	20m endurance shuttle run	cardiorespiratory endurance	cardiorespiratory endurance

[a]Anthropometric measures included height (cm) and weight (kg). Identification data included age (years, months) and sex.

Source: Modified from Eurofit Tests of Physical Fitness, Council of Europe, Committee for the Development of Sport, Rome, 1988.

were lacking that precisely described how the test must be carried out. In order to eliminate these methodological problems of comparability, a coordinated effort started in 1978 as an initiative of the Council of Europe's Committee for the Development of Sport; and concepts of a Eurofit test battery were formulated.[193]

The Eurofit test battery consists of nine fitness tests (Table 4.6). The reliability of most Eurofit tests is reportedly high.[53] This standardized test battery is in use in all European countries in order to develop population-based references for boys and girls of different age groups in each country. The Eurofit test battery has been generally recommended for use in children older than 10.[53,118,194,480,658] Eurofit tests are optionally suitable to use in children aged 6 to 10.[362,640] In 1990, the first Eurofit test results from different European countries were presented at the conference in Izmir (Turkey).[4,361,423] Some of the existing reference values for different European countries for prepubertal boys and girls are presented in Tables 4.6 and 4.7.

One of the best motor ability testing systems is utilized in Slovenia.[609] Physical characteristics and motor abilities of children are carried out by means of the following measuring procedures:

- Body stature — longitudinal body dimensionality
- Body mass — body voluminosity
- Upper arm skinfold — the amount of subcutaneous fat
- Arm plate tapping — speed of alternative movements
- Standing broad jump — explosive power
- Polygon backwards — coordination of body movements
- Sit-ups — trunk muscles strength
- Forward bend and touch on the bench — flexibility of body
- Bent arm hang — muscular endurance of the shoulder girdle and arms
- 60 meter run — sprint speed
- 600 meter run — general endurance

Following a 5-year period in which a sample of 10% of Slovenian children and youth was measured, the sports educational chart was gradually introduced in all Slovene schools from the school years 1986–1987 to 1989–1990. In the school year 1986–1987, children from the first and fifth grades of primary schools and the first grades of secondary schools were monitored; and in each consecutive year, children of the next higher class of primary and secondary schools were included. The measurements were performed by physical education teachers. The group of measurers also included other teachers and children who were specially trained for the purpose. All the measurements were performed every school year from the 1st to the 20th of April during regular physical education classes. All data were entered in the computer at the Faculty of Sport (University of Ljubljana). The faculty provided feedback in writing no later than 3 weeks after the receipt of data. The results of

Table 4.6   Eurofit Test Results in Prepubertal Boys from Different European Countries

Test Reference	Age	Belgium (18)	Estonia (313)	Lithuania (313)	Slovakia (439)	North Ireland (504)	Poland (677)	Spain (194)
20 m endurance shuttle-run (min)	9 yrs	4.6			5.35			5.58 ± 1.83
	10 yrs	5.1			5.31			6.30 ± 1.90
	11 yrs	5.7	7.4 ± 1.2	6.6 ± 1.5	5.41			7.02 ± 1.83
	12 yrs		8.2 ± 2.0	7.2 ± 1.7	5.60			
hand grip (kg)	9 yrs	17.1			22.03 ± 4.22			16.87 ± 3.10
	10 yrs	18.7			24.56 ± 4.64			18.86 ± 3.67
	11 yrs	21.0	24.0 ± 3.7	19.7 ± 4.2	27.87 ± 4.81	20 ± 4	22.0 ± 5.0	22.17 ± 4.87
	12 yrs		26.1 ± 5.6	21.7 ± 5.0	30.40 ± 5.65	23 ± 5	24.7 ± 5.5	
standing broad jump (cm)	9 yrs	143.3			149.30 ± 17.87			142.20 ± 16.96
	10 yrs	151.6			160.85 ± 18.69			148.65 ± 17.97
	11 yrs	160.7	166.5 ± 20.6	166.5 ± 17.7	167.52 ± 19.42	145 ± 19	150.9 ± 22.5	160.42 ± 19.47
	12 yrs		174.6 ± 17.8	173.4 ± 16.7	171.62 ± 18.28	150 ± 20	152.2 ± 29.0	
bent arm hang (sec)	9 yrs				19.17 ± 14.38			12.60 ± 10.20
	10 yrs				21.64 ± 17.67			14.83 ± 12.74
	11 yrs		19.0 ± 12.4	19.0 ± 12.2	25.76 ± 20.06		22.9 ± 24.1	16.67 ± 12.86
	12 yrs		23.9 ± 16.5	23.1 ± 15.4	31.15 ± 22.07		19.5 ± 20.7	
sit-ups	9 yrs	19.9			21.29 ± 5.21			
	10 yrs	21.4			23.27 ± 3.92			
	11 yrs	22.7	22.9 ± 3.5	24.0 ± 3.2	23.36 ± 4.45	22 ± 4	21.1 ± 3.8	
	12 yrs		23.7 ± 4.4	25.2 ± 3.6	24.60 ± 4.64	23 ± 4	22.7 ± 3.5	
10 x 5 m shuttle run (sec)	9 yrs	23.5			22.73 ± 2.38			17.41 ± 4.59
	10 yrs	22.8			21.17 ± 1.78			19.12 ± 5.20
	11 yrs	22.3	21.3 ± 2.1	22.2 ± 1.5	21.13 ± 2.24	21.9 ± 1.9	22.2 ± 4.0	20.18 ± 4.49
	12 yrs		20.6 ± 1.5	21.8 ± 1.4	21.10 ± 1.97	21.6 ± 1.8	23.2 ± 2.8	
plate tapping (sec)	9 yrs	18.1			16.91 ± 3.13			20.66 ± 2.14
	10 yrs	16.5			14.34 ± 2.02			20.39 ± 2.78
	11 yrs	15.1	14.5 ± 1.8	14.2 ± 1.5	13.03 ± 1.54		13.1 ± 2.5	19.54 ± 2.00
	12 yrs		13.0 ± 1.4	13.7 ± 1.6	12.93 ± 1.98		12.1 ± 2.0	
sit-and-reach (cm)	9 yrs	18.9			20.19 ± 5.54			15.10 ± 2.62
	10 yrs	18.4			18.14 ± 5.67			13.90 ± 2.06
	11 yrs	18.4	19.7 ± 5.2	18.5 ± 5.3	16.27 ± 6.30	16.5 ± 6.0	17.9 ± 5.4	12.87 ± 1.70
	12 yrs		19.4 ± 5.7	20.1 ± 5.6	15.70 ± 6.43	15.0 ± 6.5	15.7 ± 6.5	
flamingo balance (number of mistakes)	9 yrs	19.6			12.81 ± 7.41			18.32 ± 5.70
	10 yrs	17.6			11.78 ± 5.96			18.69 ± 6.15
	11 yrs	16.3	11.0 ± 5.2	12.4 ± 4.9	11.61 ± 5.31		14.9 ± 8.6	18.75 ± 5.96
	12 yrs		11.8 ± 6.2	13.0 ± 5.2	11.79 ± 5.10		12.9 ± 7.1	

Table 4.7 Eurofit Test Results in Prepubertal Girls from Different European Countries

Test Reference	Age	Belgium (18)	Estonia (313)	Lithuania (313)	Slovakia (439)	North Ireland (504)	Poland (677)	Spain (194)
20 m endurance shuttle-run (min)	9 yrs				4.57			
	10 yrs	3.5			4.94			4.48 ± 1.43
	11 yrs	3.9	6.6 ± 2.0	5.8 ± 1.4	4.82			4.97 ± 1.65
	12 yrs	4.2	6.6 ± 1.7	5.9 ± 1.4	4.85			5.36 ± 1.56
hand grip (kg)	9 yrs				20.37 ± 4.85			
	10 yrs	15.4			21.75 ± 4.26			15.68 ± 3.23
	11 yrs	17.7	20.3 ± 3.8	17.5 ± 3.9	25.42 ± 5.19	19 ± 4	17.7 ± 4.9	17.92 ± 3.77
	12 yrs	20.1	23.1 ± 5.1	19.4 ± 5.7	27.36 ± 4.81	20 ± 4	22.5 ± 5.3	21.47 ± 4.96
standing broad jump (cm)	9 yrs				140.36 ± 16.16			
	10 yrs	136.2			150.05 ± 16.63			134.28 ± 16.33
	11 yrs	145.7	157.5 ± 20.2	157.4 ± 17.6	154.52 ± 18.43	131 ± 19	135.7 ± 22.7	141.46 ± 18.62
	12 yrs	153.7	161.4 ± 20.6	161.8 ± 17.3	164.67 ± 19.87	136 ± 19	142.1 ± 24.3	147.04 ± 18.53
bent arm hang (sec)	9 yrs				8.93 ± 7.73			
	10 yrs	6.9			11.78 ± 10.22			7.72 ± 7.70
	11 yrs	6.9	10.3 ± 9.3	10.7 ± 8.4	18.61 ± 15.59		9.65 ± 8.21	8.65 ± 8.19
	12 yrs	6.8	10.4 ± 9.4	11.4 ± 8.8	17.89 ± 15.31		7.38 ± 6.08	9.71 ± 8.60
sit-ups	9 yrs				21.17 ± 4.48			
	10 yrs	18.2			21.61 ± 3.85			15.38 ± 4.86
	11 yrs	19.4	21.1 ± 4.1	23.0 ± 3.4	21.67 ± 4.31		23.32 ± 2.34	16.41 ± 5.22
	12 yrs	20.3	21.2 ± 3.8	22.5 ± 3.3	23.22 ± 4.39		23.22 ± 2.03	18.12 ± 4.67
10 × 5 m shuttle run (sec)	9 yrs				23.34 ± 1.71			
	10 yrs	24.1			22.27 ± 1.97			21.61 ± 2.39
	11 yrs	23.4	22.2 ± 2.0	22.6 ± 1.3	21.86 ± 1.98	19 ± 4	20.8 ± 4.1	20.72 ± 2.04
	12 yrs	22.9	21.9 ± 2.0	22.3 ± 1.4	21.47 ± 2.09	20 ± 4	20.9 ± 4.1	20.64 ± 1.96
plate tapping (sec)	9 yrs				16.11 ± 2.30			
	10 yrs	17.6			13.82 ± 1.80			14.79 ± 2.31
	11 yrs	15.9	13.8 ± 1.7	14.0 ± 1.4	12.89 ± 1.54	23.5 ± 1.8	12.52 ± 1.68	13.52 ± 1.70
	12 yrs	14.6	12.9 ± 1.5	13.6 ± 1.6	12.68 ± 1.57	22.6 ± 2.0	11.45 ± 1.75	12.86 ± 1.44
sit-and-reach (cm)	9 yrs				22.92 ± 5.20			
	10 yrs	22.4			21.89 ± 6.13			22.73 ± 5.34
	11 yrs	22.5	22.2 ± 7.5	21.5 ± 6.0	21.64 ± 6.20	20.5 ± 6.5	18.9 ± 6.3	23.04 ± 5.70
	12 yrs	23.1	25.0 ± 6.0	22.4 ± 5.9	22.29 ± 6.53	20.5 ± 6.0	22.0 ± 5.6	24.87 ± 6.06
flamingo balance (number of mistakes)	9 yrs				11.56 ± 7.25			
	10 yrs	17.8			10.52 ± 6.67			
	11 yrs	15.9	13.0 ± 7.8	14.1 ± 5.9	13.27 ± 5.97		12.2 ± 6.8	
	12 yrs	14.9	10.7 ± 6.6	13.0 ± 5.8	11.58 ± 5.56		11.6 ± 5.0	

measurements were assessed annually by physical education teachers in May and June. Besides annual results of measurements, the sports educational chart also included a graphic representation of physical and motor development of children over a period of several years.[609]

In Hungary, the Hungarian National Growth and Physical Fitness Study was organized in the early 1980s.[176] The testing has been carried out on a very careful sampling of 41,000 healthy boys and girls, 3 to 18 years of age, representing 1.5% of all children and youth of these ages. The study comprised a detailed anthropometric program (18 body measurements), several physical fitness tests, and data concerning the socio-demographic and cultural background of the child's family. The following motor ability tests were used:

- Hand grip
- Standing broad jump
- Medicine ball push
- Sit-up
- Burpee test
- 60-meter dash
- Cooper test

The Czech Republic has updated the Unifit test for the population ranging from 6 to 60 years of age.[425] Researchers specified standards for testing each item and overall standards expressing a personal test profile. Three tests — standing long jump, sit-ups during 60 seconds, and 12-minute run-walk (alternatively endurance shuttle-run and 2-km walk) — are recommended for ages 6 to 60. The fourth test is determined by the age of the tested person and represents that motor ability item that is characteristic and important for a particular age category (4×10 meter shuttle-run test for ages 6 to 15, pull-up test for males and flexed arm hang test for females aged 15 through 30–40, and sit-and-reach test for ages 30–40 to 60).[425]

A relatively simple test battery for testing children between 11 and 15 years of age was presented in Austria.[338] The testing battery consists of a 20-meter run, standing long jump, pull-ups, boomerang run, and 8-minute run. Several muscle function tests were also recommended. The authors noted that the tests are highly valid and reliable, and the national standard scales have also been presented.[338]

In Finland, motor ability tests for schoolchildren were presented in 1964[320] and in 1976.[449,450] In his later test battery, Nupponen[450] presented the following motor ability tests for 9- to 21-year-old people: distance run, sit-ups, flexed arm hang or pull-ups, 50-meter run, shuttle run, standing broad jump, handgrip strength, and forward trunk flexion. He presented some of the problems of his test battery: lack of long-term endurance, speed endurance, and bounce measures; low validity of girls' distance run; low reliability of shuttle run; environmental sensitivity of runs and standing broad jump; and difficulty of pull-ups in boys under the age of 16. Validity and reliability were lower in girls than in boys. His final conclusion was that fitness of Finnish schoolchildren exceeded that of school children in other countries in most fitness tasks. In his last study on motor abilities of Finnish schoolchildren at the age of 9 to 16 years, Nupponen[451] presented 14 motor ability tasks, which are too many for routine mass testing in schoolchildren. The last standard scales were presented in 1999.[452]

In Portugal,[222] a motor ability test battery was presented mostly for talent selection of boys and girls for ages 11 to 15. This test battery consisted of sit-and-reach, 50-meter dash, 2-kg ball throw, hockey ball throw, standing long jump, 10 × 5 meter shuttle run, handgrip, sit-ups, and 12-minute run. This motor ability test battery appeared to be highly reliable.[222]

Fewer test batteries for testing motor abilities of children were presented in Asia than in the U.S. and Europe. The Singapore NAPFA test battery consists of a 1600-meter run (under 12 years) or 2400-meter run (12 years and older), 60 seconds of sit-ups, pull-ups for males and flexed arm hang for females, sit-and-reach, standing long jump, and 4 × 10 meter shuttle-run.[491] All of the Singapore Ministry of Education schools administer this test battery annually.[491] In Japan,[333] vertical jump, back strength, and trunk flexion have been used for more than 10 years to test the motor abilities of children 10 years old. Physical fitness scores have been included in the physical fitness test battery.[333] In China, the motor ability test battery[621] consists of an endurance run (800 meters for children aged 10 and 11 years and 1000 meters for children aged 12 through 17 years), sit-ups in 60 seconds, pull-ups (modified pull-ups for girls), sit-and-reach, and skinfolds (sum of triceps and calf).

The quantitative assessment of physical fitness in children is one of the most complex problems in exercise science. Health-related fitness test batteries suitable for use in school environments that provide highly valid and reliable measures of exercise-induced fitness are not currently available. Fitness testing and monitoring can be valuable components of a health-related fitness program if they are used to: (1) encourage positive attitudes toward health-related fitness; (2) increase the understanding of principles underlying health-related fitness; and (3) promote a lifetime commitment to health-related fitness. The advocates of fitness test batteries often assume that tests motivate children. However, there is not enough evidence to support this assumption; and parallels in other areas of education would find supportive evidence only for those children who do well.

Few data are available about the validity and reliability of recommended motor ability test batteries. More data are available about each individual test as well as the overall validity and reliability of the test battery. The whole test battery reliability has not been mentioned. Perhaps the methodology for estimating this reliability coefficient is not readily available.[542] More than 15 years ago, Wood and Safrit[684] proposed using a modification of a multivariate statistical procedure developed by Thomson[632] to estimate the reliability of a battery of tests in physical education. A canonical correlation model was recommended using test/retest data.[684]

Only highly valid and reliable single tests were selected for the AAH-PERD test battery.[6] Safrit and Wood[542] studied the reliability of the Health-Related Physical Fitness Test battery that consisted of four subtests — 9-minute run test, skinfold measures, sit-and-reach test, and sit-up test — in 11- to 15-year-old children. The univariate reliabilities were acceptably high with the exception of the distance run test.[542] Pate et al.[474] studied the reliability of the field tests of upper body muscular strength that were used in different test batteries in 9- to 10-year-old children (pull-ups, flexed arm hang, push-ups, Vermond modified pull-ups, and New York modified pull-ups). They concluded that all tests were quite reliable (r = 0.80 to r = 0.95). The Vermond modified pull-ups test results were also reliable in the Cotton[137] investigation. Some data exist about the reliability of single tests that are part of the whole test battery. When this evidence is reviewed carefully, it is clear that the validity and reproducibility of a single test cannot automatically be generalized across age groups and genders.

Table 4.8   The Validity of Different Running Tests in Prepubertal Children

Distance	Subjects	r[a]	Reference
3/4 mile	20 boys, age 8	-0.64	Krahenbuhl et al.[347]
1 mile	20 boys, age 8	-0.71	Krahenbuhl et al.[347]
3/4 mile	18 girls, age 8	-0.22	Krahenbuhl et al.[347]
1 mile	18 girls, age 8	-0.26	Krahenbuhl et al.[347]
6 minutes	69 boys, age 9 to 12	0.50	Vodak and Wilmore[657]
12 minutes	17 boys, age 11 to 14	0.65	Maksud and Coutts[396]
1800 yards	15 boys and girls, age 11	-0.76	Gutin et al.[252]
9 minutes	22 boys, grades 1 to 6	0.82	Jackson and Coleman[297]
12 minutes	22 boys, grades 1 to 6	0.82	Jackson and Coleman[297]
9 minutes	25 girls, grades 1 to 6	0.71	Jackson and Coleman[297]
12 minutes	25 girls, grades 1 to 6	0.71	Jackson and Coleman[297]
1 mile	140 boys, age 10	-0.66	Cureton et al.[142]
1 mile	56 girls, age 10	-0.66	Cureton et al.[142]

[a]Correlations between directly measured VO_{2max}/kg (ml•min^{-1}•kg^{-1}) and distance run.

Several studies about the validity of different running tests in prepubertal children were presented more than 20 years ago (Table 4.8).[142,252,297,347,396,657] As a rule, the reliability coefficients are slightly higher in boys than in girls. However, several factors could influence the run/walk test results. For example, motivation has a significant effect upon tests involving endurance. Test results can be influenced by rewards, competition, coaction, audiences, reference standards, and/or different forms of feedback. Therefore, test conditions should be standardized as much as possible. Special problems exist with preschool children since they generally are not yet ready for long-time maximal effort. The run/walk test results reflect complex determinants. However, the cardiorespiratory function is the dominant factor that is reflected by distance running/walking performance.

Reliability estimates of the bent-knee sit-up test have been found to range from r = 0.62 to r = 0.93 in boys, from r = 0.64 to r = 0.94 in girls aged 11 to 14 years,[542] and r = 0.79 in 11- to 13- year-old boys and girls.[601] The pull-up test appears to be very reliable. In boys 11 to 13 years old, reliability coefficients have been r = 0.89.[601] One reason for the high reliability may be that so many children are unable to execute a single pull-up on either the test or retest days. The reliability of the sit-and-reach test is also high. Coefficients of r = 0.94 to r = 0.97 for 11- to 14-year-old boys[542] and r = 0.80 to r = 0.96 for girls the same age[208] have been reported. The reliability of the sit-and-reach test results by Mathews et al.[415] in third- to sixth-grade boys was relatively high (r = 0.84 to r = 0.89). Glover[234] indicated that the reliability coefficient of the sit-up test was r = 0.78 in 6- to 9-year-old girls, while it was relatively higher in boys of the same age (r = 0.91).

Presenting the validity of a single test (except endurance run tests) is more difficult than presenting the reliability. Validation results are available for upper body muscular strength tests in 9- to 10-year-old children.[474] Pull-ups, flexed arm hang, push-ups, Vermond modified pull-ups, and New York modified pull-ups tests did not validate well with laboratory strength tests. Criterion measures were performed using a supported weight system (set resistance Universal Gym) and were selected to stimulate the movements performed with the various field tests. The study found that performances on currently used field tests of upper body muscular strength and endurance are not statistically significantly correlated with laboratory measures of absolute muscular strength or muscular endurance.[474] In contrast, test performances were significantly associated with measures of muscular strength expressed relative to body mass. The observed validity coefficients were in a range of r = 0.50 to r = 0.70. Therefore, the field tests were, at best, moderately valid measures of body mass relative to muscular strength.[474] Cureton et al.[141] and Engelman and Morrow[182] also concluded that body mass was the major confounder of the strength test results in prepubertal boys and girls.

Fitness testing should be an integral part of teaching — not an isolated component of education. Teachers are encouraged to promote desirable fitness behaviors in children as opposed to the attainment of a high level of fitness.[216] Several motor ability test batteries consisting of different performance-related or health-related tests have been presented for prepubertal children. There is not one well-known test battery that is accepted everywhere. Therefore, it is difficult to compare the results of different motor ability test batteries in children of different countries. However, all presented test batteries have been recommended to measure well-known health-related motor abilities. Each child grows and develops at his or her own particular rate. It is extremely difficult to separate the contributions of growth, maturation, and exercise from any observed changes in motor performance. The use of different norm tables confuses the issue of relative fitness since the tables have been constructed on the basis of chronological age and cannot logically be used to classify individual children at different levels of biological maturation.

4.9 Motor abilities of children in different countries

Motor ability tests are usually used to motivate children to achieve higher levels of fitness and to encourage optimal levels of physical activity in their present and future lifestyles. Are boys and girls physically fit, and have their fitness levels changed over the years? Opinions of researchers vary. Some of them, based on their performance-related investigations, say that boys and girls today are more fit then several decades ago, while others say the opposite. Dinubile[166] concluded that youth fitness in the U.S. reveals some alarming trends; children are fatter, slower, and weaker than their counterparts in other developed nations. Updyke[645] suggested that physical fitness of American boys and girls has remained unchanged during the past three decades. Blair[66] concluded that youth fitness in the U.S. is exaggerated. Some researchers[488,500] have noted an erosion of youth fitness levels during the past two to three decades. Unfortunately, the detractors to this point of view say that most of the fitness comparisons have been based on motor skill tests rather than on health-related fitness tests.[500] During the past 10 years, Updyke and Willett[646] have demonstrated an approximate 10% decline in the aerobic fitness levels of children as measured by distance runs. Both boys and girls have shown marked declines. In Germany, Brandt et al.[89] recently published findings on motor development of primary school children (7 to 10 years) in 1985 and 1995 and established a distinct reduction in motor performance ability.

Normally, fitness testing in schools is conducted at the beginning of the academic year and then repeated at least once at the end of the year to determine whether fitness levels have improved, declined, or remained constant. If children are tested only once per academic year, then it is better to do it at the end of the year rather than at the beginning. The physical activities of children during long summer holidays are very different from those during the school year, and it is not recommended to test children at the beginning of the academic year who were completely passive during holidays. Reasons to test children are:

- Fitness assessments provide information to children, their parents, and teachers regarding current levels of fitness in children. This information can be reported in the form of criterion referenced. Accordingly, children will know how they have performed in comparsion with other children of the same age and sex.[674]
- Fitness assessments provide baseline data for children to set goals to improve their levels of fitness. Fitness testings increase the intrinsic motivation of children to achieve more healthy lifestyles.[150]
- Fitness test results can be utilized by teachers to gauge the effectiveness of fitness activities that have been incorporated into the physical education program over a period of time.[280]

Unfortunately, traditional fitness testing in schools is often a long and tedious process; and when physical education teachers administer each fitness test individually, the entire process may take several days. The Cooper Institute for Aerobic Research[129] offers helpful recommendations.

Body composition significantly influences physical fitness in children. Malina et al.[408] studied more than 6700 girls between 7 to 17 years of age. Adiposity was estimated as the sum of five skinfolds, and several health-related motor performance tests were used. In each age, the fattest 5% and the leanest 5% were compared in each fitness test. The fattest girls generally had poorer levels of health-related and motor fitness.[408] The level of physical fitness in children has become lower and lower, while the level of somatic growth, measured by body stature and body mass, is greater in every generation. However, some believe that physical fitness of successive generations of youth is not getting worse but rather exhibits a tendency of permanent improvement. The least-held opinion is very similar to the results of the Przeweda[487] investigation in Poland, where very large groups (more than 100,000 boys and girls) were studied in 1979 and 1989. Improved results during the decade took place in the 60-meter run, 4×10 meter shuttle-run, standing long jump, medicine ball throw, trunk bending, and sit-ups values. For hand grip strength, the results were similar for bent arm hang in girls and pull-ups in boys.[487] The level of fitness improved despite the greater effects of civilization, decreased physical activity in everyday life, and the period of crisis during the 1980s that had a negative effect on the nourishment of youths in Poland.

In a review article, Simons-Morton et al.[587] concluded that children are the most fit of all age groups. This conclusion was reached using the cardiorespiratory fitness (mostly VO_{2max}) parameters. Bar-Or[34] did not agree with this conclusion. He noted that a child has a much smaller metabolic reserve and will tire earlier than other persons. Even though children are more active than older age groups, Bar-Or[34] concluded, their fitness, in its broad sense, is lower than in young adults. Cumming et al.[139] studied the effects of increased physical education class time on improved aerobic fitness levels and found that fairly fit children are not likely to change over a school year, no matter how many hours are allotted or what facilities are available for physical education. Improving the physical fitness levels requires a training program designed to improve various fitness components.

Anyone who has done fitness testing has questioned why a person who has high scores (but who does no regular exercise) is more fit than a person who does regular exercise but has a relatively low fitness score. Cardiovascular fitness is limited in children by heredity and highly influenced by maturation.[84,348] Probably the heredity factors can stop some children's top sports careers on the endurance events but never stop in the middle nonsportsmen level.

Sex differences in physical fitness items favor boys during prepubertal ages. Environmental factors are primarily responsible for gender differences in most motor performance tasks. Factors influencing gender differences may be attributed to differing encouragement, practice opportunities, and reinforcement patterns of boys and girls inside and outside the school setting.[286] Children of the same chronological age have widely disparate biological ages, often up to 6 years.[121] Judging the test performance of children according to chronologically-based standards or norms is obviously inappropriate and could be disheartening to late-maturing boys and early-maturing girls.

An overview of results concerning sexual dimorphism in motor abilities has been presented by Malina and Bouchard.[407] As a rule, boys score higher in motor tests of strength, endurance, running, and jumping.[89,130,434] Gender differences are smaller in preschool and early elementary school years. Some authors[580,627] suggested that environmental factors are primarily responsible for gender differences in most motor performance tasks prior to puberty. For example, parents and teachers expect certain types of sex role behaviors, and boys and girls are treated as though they should perform motor tasks differently.[580] From a more powerful physiological basis, the faster skeletal maturation in girls prior to puberty suggests that their motor abilities should actually be better. Biology seems to offer little explanation for motor performances prior to puberty; differences are primarily environmentally induced.[627] The fact that it does not argues for a sociocultural explanation.[630] For example, the differences between boys and girls were smaller in AAHPERD Youth Fitness test scores when girls were given equal opportunity to learn and perform fitness activities.[253] The results of the National Children and Youth Fitness Study (NCYFS I and II) indicate that boys consistently outperform girls on all fitness measures except flexibility from ages 6 through 18.[147,525,526] The additional analysis of this data indicates that the mile run, chin-ups and sit-ups reflected similar patterns in the development of gender differences. Uncorrected effect sizes were small when children entered school.[630] However, the differences between boys and girls increased gradually during elementary school.[630] The Flemish youth study[363] using Eurofit tests indicated that sex differences were rather small between 6 and 12 years of age. However, boys obtained better results than girls at all ages for performance-related fitness tests such as static strength, explosive strength, and running speed. The growth curves were parallel between 6 and 12 years of age. No substantial sex differences were noted for the Flamingo balance test and for plate tapping.[363]

One of the first reviews about gender differences in motor performance was presented by Maccoby and Jackson[390] in 1974, and one of the latest was presented by Thomas and Thomas[628] in 1988. Excellent large-scale meta-analysis on this topic was published by Thomas and French[627] in 1985. There are small differences in the physical fitness between genders in early childhood. However, the results in our laboratory indicate that performance of six motor ability tests in boys was generally better than in girls at the ages of 4 and 5.[454]

In contrast, significant differences were found only in four tests when using the Eurofit test battery in 6-year-old boys and girls.[456,457] Girls were better in Flamingo balance and boys in handgrip strength, standing broad jump, and endurance shuttle-run.[456,457] Thomas and French[627] indicated that gender differences were small to nonexistent in 19 out of 20 motor tasks for children at 3 years of age. However, they found relatively large gender differences in throwing as early as the age of 3 years.[627] The same results were also presented in our study in 4- and 5-year-old children.[454] Heredity may be involved in the development of gender differences for certain types of activities.

There are no or very small differences between sexes across childhood in tests that characterize balance, catching, pursuit rotor tracking, tapping, and, surprisingly, vertical jump.[627] Differences between boys and girls during childhood could be explained with differing parental treatment of boys and girls of preschool age, followed by teachers in elementary school who continue to treat boys and girls differently. Enviromental factors are also important because boys participate in a wider variety of organized sports games and practice throwing than girls do.[255] However, when large groups of 6- to 12-year-old swimmers and tennis players were compared, Blanksby et al.[69] indicated that there were only very few differences between boys and girls in motor ability parameters.

Thomas and Thomas,[628] re-analyzing large groups of children, indicated that gender differences in effect sizes for mile run (half-mile run for younger children) were less than 0.5 standard deviation units until the age of 8, while effect sizes were between 0.5 and 1.0 through to the age of 12. Gender differences in chin-ups (modified pull-ups for younger children) were less than 0.5 standard deviation units in favor of boys until after the age of 9. Effect sizes were between 0.5 and 1.0 from 10 to 12 years of age. Effect sizes for sit-ups were less than 0.3 through age 8 (favoring boys) and less than 0.5 through age 12. In contrast, the sit-and-reach test performance was better in girls than in boys at all ages. Effect sizes were 0.3 to 0.6 from 6 to 11 years of age. All gender differences prior to puberty in health-related fitness test items are likely due to the different society expections for girls. Gender differences increase with age.[628]

There is only one study that has quantified comparative influences of physical growth and environment on motor performance differences in preschool children. Nelson et al.[445] reported that physical variables accounted for a significant portion of the difference. When physical variables were not considered, performance of girls was only 57% of that in boys. However, girls' relative performance increased to 69% of that in boys when throwing performance was adjusted for physical variables.[445]

Traditionally, fitness has become synonymous with aerobic or cardiorespiratory fitness, especially when it is discussed in the context of health.[34,339,348] The rationale for this tradition is that aerobic fitness affects risk for coronary artery disease. This approach, however, ignores several other components of fitness that may be relevant to health, particularly in pediatric populations.[33] These include muscle strength, muscle endurance, flexibility, and body adiposity. Kraus and Raab[350] emphasized the importance of fitness for health. Their term hypokinetic diseases referred not only to cardiovascular conditions but to many other health problems associated with sedentary lifestyles.[350]

A national survey of Australian children was initially organized in 1971.[678] A larger one was conducted in 1985[490] and the last one in 1997.[170] The 1997 results were compared with the results of 1985.[422] A marked decline in the 1.6-km run/walk and a smaller decline in the 50-meter run was found in a large group of Tasmanian children.[422] The 1997 study[170] indicated that 10- to 11-year-old children showed slower results in the 1.6-km run (by 38 to 48 seconds) and the 50-meter run compared with the 1985 study.[490] Children in the 1997 study were heavier and showed greater body mass for stature. There were few differences between the fittest and leanest quantiles in 1997 and their 1985 counterparts; but the least fit and fattest quantiles were markedly worse in 1997. This suggests that the decline in fitness of Australian schoolchildren is not homogeneous and that studies should target groups where the decline is most marked.

It is difficult to compare fitness results with different measurements since surveys often use different test protocols and analytical methods and probably reflect the social changes in the country. Sometimes it is better when children are specially prepared or trained for the test battery.[354,666] For example, Scottish schoolgirls 12 to 15 years old improved their one-mile run times by 37 to 48 seconds over three trials in the space of two weeks, presumably due to improved tactical awareness rather than a training effect.[666] Field tests are affected by environmental conditions that could be attributed to physiological stress and motivation.

A recent study applying a cross-sectional approach aimed to establish smooth curves for motor performance tests in 10- to 17-year-old Estonian girls.[383] It was concluded that girls become significantly better from the ages of 11 to 12 and 12 to 13 in all used motor ability tests, excluding standing long jump. This is surprising since Malina and Bouchard[407] have indicated that the standing long jump performance increases until 12 years of age in girls.

The China–Japanese cooperative study in physical fitness of children and youth was organized in the 1980s.[435] The subjects who participated in this study were 7 to 20 years old. Physical fitness was measured using 11 different tests. Most of them were well known (grip strength, vertical jump, 50-meter dash, etc.). The comparison of results indicated that Japanese were similar or superior to the Chinese in all items. Japanese participants outperformed Chinese counterparts in back strength, vertical jump, shuttle run, and 5-minute run. No significant differences were observed in other tests results.[435]

The Hungarian National Growth and Physical Fitness Study[175] indicated that there were fundamental differences between the performances of boys and girls in each age group (3 to 18 years of age) and each test. The differences among the age groups of girls decreased with advancing age, and they finally stabilized at relatively low values and a relatively early age. For each performance test, the performances of the age groups of girls gradually decreased and became stable at the age of 12 to 13. Somewhat surprisingly, the highest performance differences were between the ages of 8 and 9 years in boys and the ages of 7 and 8 in girls.[27] In Hungary, large differences in motor ability test results were also found between urban and rural children. The urban children were taller and heavier than rural children.[28,29]

Folson-Meek et al.[209] studied the relationships between three measures of upper body strength and endurance (pull-up, flexed arm hang, and modified pull-up) and age, body mass, percent body fat, and BMI in 104 elementary school children in grades one through six. The results indicated that the age and percent body fat were the best predictors of pull-up and flexed arm hang scores, whereas age and BMI best predicted the modified pull-up score.

Anaerobic power and capacity are usually lower in children than in adults. In children, anaerobic performance depends mainly on lean body mass and the mass of exercising muscles.[428,651] Probable reasons for the relatively low anaerobic performance in prepubertal children are:

- Limited anaerobic glycolysis in children[189]
- Relatively low phosphofructokinase activity in muscle[188]
- Lower acidosis in blood following exercise[414]
- Possible effect of lower testosterone concentration in blood[202]

The factors influencing strength results before puberty are different and partly contradictory. The concentration of testosterone in blood before puberty is low in boys. However, the percentage of muscle mass increases to the same extent as it does between puberty and maturity.[430] The increase in secretion of testosterone at puberty has been associated with increases in skeletal muscle mass.[430] In a longitudinal study, Mero et al.[429] reported a significant positive relationship between testosterone concentration in blood and strength in boys 11 to 12 years old. Additionally, strength training increased the testosterone concentration only in an exercising group.[429] Thus, the responsiveness of muscle to training is not solely dependent upon the level of testosterone.

4.10 Tracking motor ability

Tracking motor ability parameters from childhood to adulthood is important because the level of motor development is influenced by several risk factors for cardiovascular and other chronic diseases. The extent to which physical fitness tracks during childhood and from childhood into adulthood and the factors that explain changes in fitness during childhood (a shift from one fitness track to another) are poorly understood. Longitudinal data are needed to understand and estimate the degree of tracking health-related fitness characteristics and how they relate to tracking of cardiovascular health outcomes during childhood. Particular attention should be paid to critical periods of growth and development — for example, the transition from early to middle childhood, pubertal transition, and transition from late adolescence to early adulthood.[31,236]

Interage correlations for several measures of muscular strength and endurance in prepubertal children are presented in Table 4.9. As a rule, correlations range from low to moderately high, and differences between boys and girls are not consistent. Harlan et al.[260] indicate that the stability of static strength and muscular endurance varies among tasks. Relationships taken at intervals of 5 to 6 years during childhood (between 7 and 12 years) range from low to moderate.[494] The study of Ellis et al.[178] tracked boys from 10 to 16 years of age and found interage correlations for sit-up scores of r = 0.40. Pre- and postpubertal children were compared in this study.[178]

Table 4.9   Interage Correlations for Several Measures of Muscular Strength and Endurance in Prepubertal Children

Reference	Span, Years	Males	Females
Hand grip			
Clarke[120]	7 to 12	0.40	
Szopa[617]	7 to 14	0.44	0.35
Composite measures			
Rarick and Smoll[494]			
Upper body[a]	7 to 12	0.35	0.26
Lower body[a]	7 to 12	0.40	0.52
Clarke[120]			
Lower body[b]	7 to 12	0.45	
Total body[c]	7 to 12	0.72	
Flexed Arm Hang			
Branta et al.[190]	5 to 10	0.34	0.24

[a]Average correlations based on Fisher's z-transformations of interage correlations for four cable tensitometric strength tests.
[b]Average correlations based on Fisher's z-transformations of interage correlations for three lower body cable tensiometric strength tests.
[c]The average of 11 cable tensiometric strength tests modified from Malina, R. M.
Source: Res. Q. Exercise Sport 67, S 48, 1996.

Interage correlations for several measures of motor fitness in prepubertal children are presented in Table 4.10. Relationships for tests of jumping and running are relatively variable among studies, vary in the interval considered, and range from low to moderately high. Instability in correlations spanning between ages 5 and 10 probably reflects variation in attainment of mature movement patterns.[405] Mature patterns of fundamental skills are not attained by some children until 8 or 9 years of age.[570] Tracking of flexibility in children has mainly targeted field measures of hamstring flexibility (the sit-and-reach test). Interage correlations are low to moderate between 5 and 10 and between 8 and 14 years of age — $r = 0.26$ and $r = 0.52$, respectively in girls, and $r = 0.36$ and $r = 0.52$, respectively in boys.[90,269]

Table 4.10   Interage Correlations for Several Measures of
Motor Fitness in Prepubertal Children

Reference	Span, years	Males	Females
	Standing long jump		
Glasgow and Kruse[233]	6 to 12		0.74
Rarick and Smoll[494]	7 to 12	0.48	0.71
Keogh[330]	6 to 9	0.60	0.70
Keogh[330]	8 to 11	0.73	0.59
Branta et al.[90]	5 to 10	0.46	0.38
	Vertical jump		
Branta et al.[90]	5 to 10	0.43	0.31
	Dashes		
Glasgow and Kruse[233]	6 to 12		0.70
Rarick and Smoll[494]	7 to 12	0.39	0.92
Branta et al.[90]	5 to 10	0.52	0.16
	Shuttle-runs		
Branta et al.[90]	5 to 10	0.24	0.46

Scarce data is available about the tracking of VO_{2max} during the prepubertal period. Only Janz and Mahoney[304] found that peak VO_{2max} tracked significantly ($r = 0.70$ to $r = 0.75$) during their three-year study of children ages 7 to 12 years of baseline. In another study[617] that spans from childhood into adolescence (7 to 14 years of age), low interage correlations were found for boys ($r = 0.24$) and girls ($r = 0.21$). Aerobic power tracks at a relatively low level ($r = 0.30$) in boys from childhood to adulthood (between 11 to 18 years of age).[604]

Several longitudinal studies have analyzed age-, growth-, or maturity-associated changes in VO_{2max} by correcting data for a single body size indicator within each analysis — usually body mass[59] or anthropometrically predicted body mass.[536] However, a more comprehensive understanding of developmental changes in VO_{2max} should simultaneously investigate the influence of other covariates. For example, despite valid concerns regarding issues of colinearity among covariates,[37] stature has been shown to be a significant, independent predictor of VO_{2max} in young people when incorporated alongside body mass in an allometric analysis.[18]

Few studies exist about the tracking of physical fitness during the pre-pubertal time. However, the existing results indicate that different measures of performance- and health-related physical fitness track significantly during childhood. Therefore, correlations are only at low to moderate levels.

4.11 General considerations

The testing of motor abilities in children is one of the oldest questions in sports science. Fitness testing is important as a part of a comprehensive health-related fitness curriculum to teach our children about the health-related benefits of exercise during prepubertal years. The selection of motor ability test batteries to test health-related physical fitness is complicated for children of all ages. Fitness tests that are suitable for use in a school environment and that provide valid and objective measures of fitness are simply not available. Fitness tests determine the obvious at best, only distinguishing the more mature and/or motivated children from the less mature and/or motivated children. Prepubertal children are not yet ready for overexertion. Fitness tests that require a long time or are painful and uncomfortable are not acceptable for prepubertal children.

The selected motor ability tests should be as simple as possible, easily understandable for children and teachers, not time consuming, and useable during physical education classes. Our selected motor ability tests for prepubertal children are presented in Appendix 4. This motor ability test battery is only a sample for possible testing and is not validated by us. However, other investigations have shown that these motor ability tests are valid and reliable.[541]

What to do with obtained motor ability test results is another question. Interpreting these results requires the help of physical education specialists, epidemiologists, and school physicians. Standardized testing results are needed to compare obtained results both with results of motor ability tests of different countries and different regions inside a country. Furthermore, is it correct to grade the results of motor ability tests in physical education classes? It is probably motivating for children to increase their level of physical fitness to obtain better results and better grades in the future. However, bad grades are stressful for children with low results of motor ability tests. Another question is the difference in biological maturation between children of the same chronological age. The use of standard scales is not stimulating for children who are exercising in sport clubs practically every day.

Anthropometrical parameters of children should also be considered when assessing the results of different motor ability tests since the anthropometrical profiles at the same chronological and biological age of children may be very different. For example, the results of motor ability tests that need speed and strength are lower in children with a smaller stature. In contrast, the results of motor ability tests that need endurance are relatively low in children who are very tall and present a relatively high body mass.

Tracking motor abilities from early childhood to adulthood is important because the level of motor development is influenced by critical and sensitive periods in ontogenetic development. Particular attention should be paid to critical periods of growth and development. It is likely that related critical events trigger the acceleration of improvement in different motor abilities — for example, the transition from prepubertal years to puberty. Different motor ability parameters track significantly during prepubertal years. However, correlations are only at low to moderate levels.

chapter five

Motor skills of prepubertal children

5.1   Introduction

Motor skill development is defined as the changes in motor skill behavior over time and the processes that underlie these changes.[1,123] *Product* (what are the changes in motor behavior) and *process* (how and why changes in motor behavior occur) are important distinctions in studying motor skills during childhood.[90,123] Many researchers have studied the developmental aspects of various motor skills during childhood and adolescence. However, little information is available about the extent to which various factors actually influence motor skill development during childhood.[402] Understanding the interaction between different genetic factors and environmental processes in the development of specific motor skills during growing years is important.[90]

Many factors influence motor skill development at a particular time period or age — including biological variables such as physical growth and biological maturation, and environmental factors such as socioeconomic status and habitual physical activity. This chapter is focused on the relationships between motor skills and physical activity, motor ability and somatic development, and motor skill development in prepubertal children.

5.2   Basic motor skills

The basic skills of running, jumping, throwing, and ball handling are of primary importance in different physical education programs. Fundamental motor skills are common motor activities with specific observable patterns that form the basis for the more specific and complicated sports and movement skills common to our culture.[226,676] Each fundamental motor skill has definable characteristics that are observable and that underlie the unique characteristics of the skill.[97,508,676] Most skills used in sports are advanced versions of fundamental motor skills.

Fundamental motor skills can be divided into three basic categories:[97,555]

- Locomotor skills — walking, running, jumping, hopping
- Non-manipulating skills — turning, balancing, sliding, leaping
- Manipulative skills — kicking, throwing, catching, striking, bouncing, pulling, pushing

The first studies on fundamental motor skills were presented about 70 years ago. The studies of Bayley,[45] Ames,[12] and Gesell and Thompson[230] recorded motor achievements of very young children in order to establish normative ages or percentile of performance. Specific behaviors were recorded according to the chronological ages of the children and their order of appearance in the movement repertoire. The general concept of these early studies was based on developmental stages of fundamental motor skills, which implies a universal, invariant sequence of motor skill development that generalizes across similar tasks.[97,509,510] Specifically, changes in movement patterns by all individual performers should follow the same sequence and should be the same for similar tasks such as jumping, hopping, throwing and striking.[97,509] For example, Gesell and Thompson[230] described up to 58 stages of behaviors for 40 different tasks. However, empirical evidence to support the validity of the concept of stages for the development of motor skills is limited.[97] Roberton[509] suggested that the levels of developmental skill progression should be referred to as steps rather than stages. Despite these shortcomings, this line of investigation provided initial evidence for the concept of developmental motor skills sequences.[12,45,97,509]

Understanding skill acquisition requires a knowledge of how the motor system is controlled as well as the processes underlying change from immature or unskilled to skilled performance. One of the pioneer studies about motor skills in children was conducted by Hellebrant et al.[276] They presented careful biomechanical descriptions of how children moved, and they documented changing patterns of motor coordination for fundamental motor skills such as jumping. Rhetoric concerning the underlying process of development was conspicuously absent.[276]

In the last 20 to 30 years we have seen a theoretical shift in the concept of motor control and coordination. In preliminary studies, motor skills were measured in quantitative terms, with little concern for the relative proficiency with which specific tasks in a progression were accomplished. However, several researchers[225,511] have emphasized that the identification of qualitative differences in measurement patterns provides a more detailed means of assessing motor developmental status than the earlier developmental scales. Branta et al.[90] state that the move to qualitative assessment of movement patterns has raised several new questions, the foremost of which is how the patterns are arranged in developmental order by researchers. They concluded that much of the research in the early 1980s described the quantitative changes (product scores) and qualitative changes (process or movement patterns) in fundamental motor skills throughout childhood and adolescence.[90] Over the decades of research, two main approaches to describe movement patterns have emerged:[90,97]

- A total body configuration approach
- A body component approach

The total body configuration approach is reportedly the simplest way to describe a particular developmental task.[90,570] Branta et al.[90] argued that there is sufficient cohesion among certain characteristics of a movement pattern to define those as stages of development. The progression from stage to stage does not imply abrupt change but rather a continuum of development, with consolidation among characteristics around a point on the continuum.[570] Seefeldt and Haubenstricker[570] have described the total body configuration stages of a wide range of fundamental motor skills on the basis of data from a longitudinal study. The ages at which 60% of the children were able to perform the various developmental levels of eight fundamental motor skills were reported.[570] For boys, the highest developmental level was first achieved for running (4 years), followed by throwing (5 years), skipping (6.5 years), catching (7 years), kicking (7 years), striking (7 years), hopping (7.5 years), and jumping (9.5 years). For girls, the highest developmental level was first also achieved for running (5 years), followed by skipping (6 years), catching (6.5 years), hopping (7 years), kicking (8 years), striking (8.5 years), throwing (8.5 years), and jumping (10 years).[570] Clark and Whitall[122] suggested that the developmental ordering of locomotor patterns is walk, run, gallop, hop and skip.

The body component approach of developmental sequence, promoted by Roberton,[506,507] is used by many other researchers to study fundamental motor skills.[256,272,431,680] Roberton[507] rejected the use of total body descriptions of motor development as inadequate and misleading. She stated that motor skills are more accurately classified according to intratask components, because stages exist at the component level only and not as total body configurations. Stage descriptions should address various components of the task separately; the stages described by total body configurations are too general and mask variability in the development of specific body components.[507] In her study, Roberton[506] used two sets of body component categories to describe the overhand throw for force (one for arm action and the other for pelvic spinal action). She found that the development of these components appeared to occur at different rates. Following Roberton's research, other researchers have presented component sequences for forward rolling,[680] hopping,[256] overarm striking,[431] and standing long jump.[272] In addition, component sequences have been hypothesized for catching, punting, running, and walking.[97]

An excellent example is presented by Haywood (1981)[272] about the qualitative changes from an early to a more mature pattern of a standing long jump skill. It demonstrated that skill development corresponded to sequential changes in some of the body components over time and with practice.[198,272] Observations of clear sequences have often led to the delineation of particular stages of skill acquisition. For example, Strohmeyer et al.[610] described components of two-handed catch (arm preparation, arm reception, hand, and body action). Two of these components (hand and body action) were found to be age related.[610] Clear developmental ordering has also been observed for arm and sequences in the forward roll between 5 and 9 years of age.[680] In children, overhand throw has been analyzed to present five distinct parts:[285,671]

- Side-on stance
- Arm extension to the rear prior to throwing
- Forward step into the foot opposite the throwing arm
- Rotation of trunk
- Complete follow through

Branta et al.[90] indicated that most children demonstrate mature or adult-like movement patterns at the preschool and primary school age (5 to 7 years), but a great variability appears in motor skills at specific ages. Age trends are available in the development of movement patterns on the basis of running,[213] jumping,[511] throwing,[445,631] walking,[174] and catching.[384] According to Gabbard[224] and Gallahue,[225] the time period from ages 2 to 7 is termed a fundamental movement phase of motor development. Sanders[555] indicated that children develop fundamental motor patterns between the ages of 2 and 7. An intensive development of motor skills occurs during the prepubertal time. High levels of motor skills were found at 11 to 12 years of age in tests of precise performance in a complicated situation, for skills of fast and precise movements, and for skills of feet.[90,654] However, the skills of hands reached the highest level in 13 to 14 years of age in girls — in most cases at the middle of puberty.[90,654]

The development of fundamental motor skills among children through quality of physical education is a potentially important contributor to successful and satisfying participation in sports and other health-related physical activities. Fundamental motor skills are motor activities with specific observable patterns and are prerequisites to the advanced skills employed in competitive sports, different games, dance, gymnastics and other physical activities.[98,227] However, standardized, widely accepted and valid tests are not yet available for some fundamental motor skills. A great practical need exists to study motor skills. For example, physical education teachers are interested in understanding how to improve the instruction of motor skills through a knowledge of motor performance changes. However, the overwhelming sentiment is that motor development researchers do what they have to do to improve the quality of life for children. They apply what is learned in elementary physical education and youth sport, and they develop special programs for children to prevent or treat motor deficits. Most of the information generated in motor development is directed at instructional programs, some at treatment programs, and the smallest portion at theoretical or medical programs.[629]

Most children have the potential to be mechanically efficient and coordinated in fundamental motor skills by the time they are 5 to 7 years old.[97,226] This coincides approximately with the age at which children enter the first grade of primary school. Children normally develop motor skills in a sequential and orderly manner.[90,97,493] Children at the fundamental motor skill stage are building upon previously learned movements and preparing for the acquisition of more advanced motor skills.[662] However, Scott[567] noted that, prior to 7 to 8 years of age, children cannot perform tasks requiring much coordination; therefore, early instruction will result in poor performance or total failure. In contrast, it has been suggested that preschool and early elementary school years are the best times for children to learn and begin to define motor skills.[90,97,226,493]

Kelly et al.[324] found that children who received a physical education program from qualified teachers performed significantly better in fundamental motor skills than did children who received supervised activity time only. The level of motor skills improves with age.[90,190,508] However, the development of motor skills does not occur automatically but is under environmental and genetic influences.[90,226,398,407] As a child gets older, environment begins to play a greater role in motor skill development.[407] Environmental factors include opportunities to practice, interest in the child's activities shown by significant others, and the quality of instruction provided.[226,592]

In preschool and early primary school years, fewer competing activities allow children more time to concentrate on developing motor skills. However, early detection of motor problems and start of intervention programs can eliminate or minimize many physical and related emotional problems.[257,600] The information available at the ages and stages in the development of fundamental motor skills is of potentially great value to teachers of motor skills, especially at the primary school level, for at least three reasons:

1. It may provide a reasonably objective method for monitoring motor development of individual children and for detecting any potential movement problems.
2. It may provide the teacher with a guide as to forthcoming progression in movement sequence development and, therefore, may provide a basis for accelerating acquisition of specific motor skills.
3. Assessment of fundamental motor skills can provide an indication of the readiness of children for involvement in more structured activities such as sports, where performance is based around proficiency in one or more fundamental motor skills.

Gallahue[226] and Seefeldt[569] indicated that limited competence in fundamental motor skills at an early age can negatively impact future performance in physical and motor activities. Ulrich[644] has reported levels of motor competence to be significantly related to participation in organized sports programs. It would be difficult to conceive that children could experience success in games and sports incorporating elementary skills. In a Wankel and Pabich[663] investigation, many children indicated that they dropped out of sports because they could not perform skills well enough to play the sport with success. Evans and Roberts[195] reported that children gain peer acceptance by excelling at something valued by other children, and there is much evidence to show that sports skills are valued by children. Robertson[513] found that 18% of boys and 24% of girls had dropped out of their favorite sport by the age of 12. It would be interesting to know the role of a low level of fundamental motor skills. Several researchers[285,324,644] have revealed that many children were unable to demonstrate mature motor skill patterns by the end of the third grade.

The development of fundamental motor skills is considered by many as a key objective of physical education programs because it increases the options for participation in games, sports, and other physical activities.[242] Haubenstricker and Seefeldt[269] identified the prepubertal period as of particular importance for the acquisition and development of motor skills in the growing child. Gallahue[225] suggested that four factors regulate the development of fundamental movement patterns — maturation, physical development, hereditary factors, and environmental experiences.

Parental involvement is among sociocultural influences in motor skill development. Greendorfer and Lewko[245] concluded that fathers have an especially important influence on sport participation in children, and fathers tend to encourage boys more than girls to participate in sports. Girls are shunted away, especially from activities that are perceived as dangerous.

Simple observation methods have frequently been used to characterize motor skill development in children.[79,242,285,662] The qualitative components of different motor skills are assessed by members of the field team by scoring each of the components as present or absent in four out of five trials.[285] That is, when children demonstrated the skill component in four out of five trials, they were recorded as possessing that skill component. Usually, evaluators who have previously trained together achieve an accuracy and interrater agreement score of greater than 90% for each skill during training.[242,662] This simple method has been used, for example, in a large study in Australian children.[79]

Recently, Booth et al.[79] studied six fundamental motor skills (run, vertical jump, catch, overhand throw, forehand strike, and kick) in a large group of Australian children (n = 5518) at the ages of 9.3, 11.3, 13.3 and 15.3 years. The findings of this study indicated that the prevalence of mastery and near mastery of each of the fundamental motor skills was generally low. There were no differences between children from urban or rural schools, and the prevalence of skill mastery was directly associated with socioeconomic status more consistently among girls than among boys.[79] In a later study of Australian children at grades two, four, six and eight (n = 1182), five motor skills that are fundamental to performance in a wide range of physical and sport activities were studied using videotaping.[662] The studied motor skills were overarm throwing, catching, forehand striking, two-hand side-arm striking, and instep kicking. The findings of this investigation were basically similar for all five motor skills. The percentage of children who achieved mature movements was alarmingly low. At the age when involvement in organized sports is likely to first take place, less than a half of the boys and less than a quarter of the girls had mastery over the fundamental skills on which most sports are based.[662]

There have been a few longitudinal studies investigating changes in motor skills in prepubertal children. For example, Roberton and Halverson[512] studied the relationships between developmental levels within five body components (forearm, humerus, trunk, stride, step) and a product variable, horizontal ball velocity, in children between 6 to 13 years of age over a 7-year period. As the children matured with age, relationships between a given component and ball velocity changed. Moreover, the relationships were also changing between components during a 7-year period.[512] This raises the question of what movement changes in one part of the body did to movements in other parts of the body. Numerous transactions must occur in the simple system of body part development over 7 years.[512]

5.3 Relationships among motor skills, physical activities, motor abilities, and somatic development

Several researchers have indicated that childhood experiences play a significant role in the future physical activity habits of adolescents and adults.[90,306,443] However, there is an opportunity to play and explore different movement activities, with little emphasis placed on the development of basic movement skills in early childhood. This is probably one reason why young people and adults do not want to return to physical activities. Their low level of basic motor skills is a serious reason. Jess et al.[306] emphasized that there are three interrelated developmental mechanisms that strongly influence the participation process: (1) basic movement skill development; (2) perceived movement competence; and (3) the role of significant others. The acquisition of basic movement skills in a developmentally appropriate manner is the most effective route to future participation.

Laws[360] also emphasized that the acquisition of movement competence is central to a physically active life. Elementary movement skills are the basis for future performance and involvement in the more specialized games, sports, dance and different recreational activities.[17] From a scientific point of view, few studies define which movements need to be developed in childhood to ensure that adults have an appropriate underpinning for a physically active life. Several researchers indicate that most children do not receive appropriate movement opportunities to develop adequate levels of elementary movement skills.[525]

Propulsive movement such as throwing is used to move objects away from the body. In terms of physical education, no doubt the most important propulsive skill is throwing because the overhand throwing pattern is assimilated into a wide variety of sport skills such as spinning a volleyball or the overhand clear in badminton. Prepubertal children at the age of 6.5 years are reportedly capable of throwing with a forward step of the opposite leg and greater trunk rotation.[358] Associated with advancing age is a tendency for children to use preparatory movements. The most advanced preparatory sequence in throwing involves a circular action in which the arm moves down and back.[358]

Investigations in our laboratory studied the relationships of physical activity and somatic growth with overhand throwing development and physical fitness in prepubertal boys and girls.[498,499] Overhand throwing was performed in field conditions as an indicator of motor skill development. Each child performed three trials from the standing position with a tennis ball (weight 150 g). The task was to throw the tennis ball as far as possible. Three trials for each child were recorded, with a Panasonic® videocamera located to the left of the child. The camera objective was placed perpendicular with the throwing direction, and the distance between camera and child was 20 meters. The distance that the ball was thrown was recorded as a quantitative measure of overhead throwing performance. The qualitative evaluation of throwing performance was done by visual observation of the videotapes. Three specialists of motor development served as observers. Each specialist had previous training with the evaluation of total body developmental sequences to assess fundamental motor skills.[268] Observers rated videos individually. Scores for each subject in each trial ranged from one to four, corresponding to stages one through four, (immature to mature throwing patterns.)[498,499]

Table 5.1   Zero-Order Correlations between Selected Somatic Characteristics and Throwing Result and Throwing Stage in Prepubertal Girls

Age groups	Throwing result				Throwing stage			
	7 n = 48	8 n = 54	9 n = 57	10 n = 56	7 n = 48	8 n = 54	9 n = 57	10 n = 56
Stature	0.22	0.29	0.29	0.31[a]	-0.13	0.20	-0.08	0.23
Body mass	-0.05	0.19	0.14	-0.08	-0.08	-0.11	0.14	-0.09
Femur width	0.19	0.28[a]	0.38[a]	0.21	0.18	-0.03	-0.19	-0.04
SSF[b]	-0.12	0.09	-0.19	-0.09	-0.14	-0.22	-0.12	0.22

[a]Statistically significant — p<0.05.
[b]SSF — sum of triceps, biceps, subscapular, abdominal, and medial calf skinfolds.
Source: Compiled from Raudsepp, L., and Jürimäe, T., *Am. J. Hum. Biol.,* 9, 513, 1997.

The correlation analysis indicated that, from somatic characteristics, the throwing result correlated with body stature in 10-year-olds and with femur width in 8- and 9-year-old age groups of prepubertal girls (Table 5.1).[499] The throwing stage does not depend on the anthropometric parameters in prepubertal girls. Physical activity, estimated by a modified 7-day physical activity recall of Godin and Shephard,[235] was also not significantly related to throwing result and throwing stage, except throwing results with moderate to vigorous physical activity in the 10-year-old group (r = 0.33) (Table 5.2). In contrast, Butcher and Eaton[99] concluded that there are close relationships between physical activity and the developmental level of several fundamental motor skills (running, throwing, and jumping). The main conclusion of our study was that the relationships between somatic characteristics and physical activity with quantitative and qualitative variables of overhand throwing performance in prepubertal girls are not significant as a rule.[499]

Table 5.2 Zero-Order Correlations between Physical Activity and Throwing Result and Throwing Stage in Prepubertal Girls

	Age (yrs)	n	TPA[a]	MVPA[b]	LPA[c]
Throwing result	7	48	-0.12	0.22	-0.06
	8	54	0.14	0.23	0.10
	9	57	-0.19	0.13	-0.09
	10	56	0.18	0.33[d]	0.06
Throwing stage	7	48	0.13	-0.06	0.03
	8	54	-0.11	0.23	0.10
	9	57	-0.16	0.23	0.04
	10	56	0.18	-0.16	0.09

[a]TPA — total physical activity.
[b]MVPA — moderate to vigorous physical activity.
[c]LPA — low physical activity.
[d]Statistically significant — $p < 0.05$.
Source: Modified from Raudsepp, L., and Jürimäe, T., *Am. J. Hum. Biol., 9*, 513, 1997.

The findings of this investigation are partially consistent with previous studies analyzing biological factors related to throwing performance in children.[90,99,446] In attempting to divide gender differences in overhand throwing into biological and environmental factors in 5- to 6-year-old children, Nelson et al.[445] found that only two somatic variables (estimated leg muscle and shoulder/hip ratio) significantly predicted throwing performance in boys, accounting for 18% of the total variance. For girls, both biological and environmental variables were significant predictors ($R^2 = 0.48$) of throwing performance.[445] Furthermore, the findings of this study indicated that, even when the results of throwing were adjusted for biological factors, the distance girls threw only increased from 57 to 69% of the boys' throws.[445] In a review article, Branta et al.[90] suggested that the fact that throwing performance tended to be stable across longitudinal investigations pointed out the biological influence on this motor skill in children. However, different environmental factors need to be considered when analyzing the correlates of motor skill development in children.[407,631]

In another study in our laboratory, relationships among throwing skills, anthropometry, and physical activity were investigated in 203 boys aged from 7 to 10 years.[498] The results indicated that the throwing result was significantly correlated with several somatic dimensions (stature, femur width); but throwing stage was not significantly correlated with somatic measures (Table 5.3). Only moderate to vigorous physical activity from physical activity parameters correlated significantly with the throwing result ($r = 0.20$). Partial correlation analysis, after removing age and moderate to vigorous physical activity, indicated that the throwing result remained significant with skeletal width.[498]

Table 5.3   Relationships between Selected Somatic Characteristics and Throwing Result and Throwing Stage in Prepubertal Boys (n = 203)

	Throwing result		Throwing stage	
	Zero-Order Correlations	Partial Correlations[a]	Zero-Order Correlations	Partial Correlations[a]
Stature	0.22[c]	0.10	-0.11	0.06
Body mass	0.12	-0.04	0.06	-0.15
Femur width	0.26[c]	0.20[c]	0.03	-0.09
SSF[b]	-0.13	-0.07	-0.10	0.13

[a]Partial correlations — controlling for age and moderate to vigorous physical activity
[b]SSF — sum of triceps, biceps, subscapular, abdominal, and medial calf skinfolds.
[c]Statistically significant — $p < 0.05$.
Source: Modified from Raudsepp, L., and Jürimäe, T., *Biol. Sport.*, 13, 279, 1996.

While several studies have investigated the association between somatic variables and quantitative measures of fundamental motor skills in children,[445,584,631] only a few have focused on the relationship between somatic growth and body composition or qualitative development of motor skill patterns in children.[446] The results of our laboratory investigations[498,499] have clearly demonstrated that the qualitative development of overhand throwing assessed by developmental stage in prepubertal boys and girls was not influenced by the somatic growth and body fat of children. Correlations found between developmental stages and somatic dimensions were consistently low in both boys and girls.[488,489] These findings indicated that the developmental level of acquisition of motor skills in prepubertal children was not related to the quantitative measures of somatic growth and body composition. Accordingly, it is expected that several sociocultural factors as well as specific physical activities related with the performance of skill-specific movements affect to a greater extent the qualitative form of motor skill patterns in prepubertal boys and girls.

The lack of significant associations between somatic characteristics and developmental stage of overhand throwing suggests that factors other than physical dimensions affect the acquisition of this skill in children. Previous studies have clearly demonstrated that environmental factors influence the development of fundamental motor skills in children.[173,255,446,564] Schnabl-Dickey[564] analyzed the effects of sociocultural factors on the throwing performance in preschool children. The results of this investigation showed that a permissive home environment was associated with superior throwing skills.[564] Opportunities for children to participate in sports, especially in sports games and events related to throwing, are very important.[353] The study in our laboratory[499] demonstrated that partial correlations (controlling for somatic characteristics) between physical activity and quantitative and qualitative measures of throwing performance were not significant. One possible explanation for the relatively low influence of physical activity on throwing performance includes the generally low participation level of girls in different throwing activities. In addition, the physical activity recall[235] used in this study did not specifically focus on the throwing activities but only on the overall level of physical activity described by intensity categories. However, the developmental level of throwing as assessed by qualitative (attained stage) and quantitative measures (distance throw) clearly indicates nonsignificant improvement in these indicators between the ages of 7 to 10 years. The lack of practice and encouragement of girls to perform throwing activities may be the primary reason for the relatively low developmental level of this motor skill by the age of 10.

Balance is very important to fundamental motor skill development.[97,643] Balance, which may be characterized as static (stationary body) or dynamic (while moving), comes into play during throwing as the child is forcefully moved from the perpendicular during vigorous forward arm action. Butterfield and Loovis[100] studied the contributions of age, sex, balance, and sport participation to the development of throwing in children aged 4 to 14 years. They assessed the throwing development by Ohio State University Scale of Intragross Motor Assessment (OSU-SIGMA).[384] Boys had more mature throwing patterns at all grades. Age within grades had minimal effect on throwing development. After the elimination of age and balance as potential predictors of mature throwing behavior, it was found that the development of throwing skill is largely influenced by gender and opportunities to participate in community and school youth sports.[384]

Examining gender differences in motor skill development is important in many practical settings, especially in the fields of physical education and youth sports. Gender differences in the performance of fundamental motor skills have been confirmed in many studies.[97] Deach[149] reported more than 50 years ago that boys were approximately one year in advance of girls in their quality of motor performance (pattern development) and showed greater ability to move with an integrated body pattern during throwing, catching, kicking and striking tasks. There are gender differences in both perceptual and gross motor skills. These differences could be the result of early sexual differentiation of the brain.[198]

As a rule, girls are better than boys in fine movement tasks. For example, girls are faster at alternate dot circle,[128] tapping,[97] bead stringing,[584] typing letters,[97] or in a serial choice response task,[97] in which they have a shorter response time than boys.[199] In contrast, boys usually perform better than girls in gross motor tasks. The difference, although sometimes present as early as preschool or elementary school years, tends to increase with age. For example, boys were found to be better than girls at keeping an arm flexed while hanging on a bar with no support, or at jump-and-reach after 8 years of age — whereas no or only slight gender differences were found in 5- to 8-year-old children.[97] Concerning running, gender differences in favor of boys were found at the kindergarten age[434,440] or at the age of 7 years.[584] Boys were also better in throwing,[440,570,584] jumping,[434,584] kicking a ball,[169] and ball catching[295] by the age of 7. The gap between sexes increases throughout elementary school in the overarm throw.[255] These differences can be primarily explained with specific gender-oriented activities. For example, ball throwing and kicking are associated with male-dominated games such as soccer. As a rule, girls were found to be better than boys in flexibility,[90] balancing,[440] or at hopping, skipping, and/or rope skipping.[93] These differences could also be explained by typical games of girls in a social surrounding. On the other hand, when adult expectations for motor ability differ, girls are generally provided with less (or lower level) instruction, encouragement, and opportunity for practice and performance of their motor skills. Subsequently, their personal expectations are generally lowered.

A meta-analysis in gender differences that summarized the existing literature of movement performance from 1965 to 1982 was performed by Thomas and French.[627] Sixty-four studies were selected that met task, subject, design, and statistical treatment requirements of the meta-analysis. The analyzed studies included more than 30,000 subjects ranging from 3 to 20 years of age, with 51% boys and 49% girls. Effect sizes were used to standardize mean differences in performance outcomes between boys and girls across the studies. Effect sizes were calculated by dividing the mean performance differences by the overall standard deviation. The effect sizes approximating 0.2, 0.5 and 0.8 were categorized as small, moderate, and large differences, respectively.[124] Fundamental motor skills of running, jumping, catching, and throwing were analyzed. For running for maximum speed over short distances and jumping for maximum horizontal distance, the effect sizes were about 0.4 to 0.5 until about 12 years of age, then increased beyond 2.0 by 18 years of age. For catching success, the effect sizes were below 0.4 until the age of 10, then increased up to 1.0 by 13 years of age. The effect sizes for throwing for maximum distance were over 1.0 between the ages of 3 to 6 years and increased beyond 3.5 by 17 years of age.[627] The results of meta-analysis demonstrated that performance differences between boys and girls in these motor skills (except throwing) were low to moderate before puberty, while after puberty differences were so large that the lowest boys outperformed almost all girls.[97,627]

Thomas and French[627] suggested that environmental influences are primary contributors to gender differences in performance prior to puberty in balance, catch, dash, grip strength, long jump, pursuit rotor tracking, shuttle run, sit-ups, tapping, and vertical jump. However, rapid increases after puberty are due to both environmental and biological factors. They also suggested that the early large differences in throwing are probably caused by biological factors, and that the expanding gap in later years is caused by both environment and biology.[627]

Seefeldt and Haubenstricker[570] classified several fundamental motor skills into developmental patterns of movement or stages that are observable as a child gains proficiency in performance of a movement task. They examined the relationships between age, stage and gender. The results indicated that in four of eight skills (catch, run, hop, and skip), girls were ahead of boys in the initial appearance (stage one) of the skills. There were not any differences between genders in the emergence of the other four skills — throw, strike, kick, and jump. Despite what appears to be a head start in girls' performances, boys preceded girls in the attainment of the most mature developmental stage in five of eight skills (throw, strike, kick, run, and jump). All differences were significant except those reported for the first stage of catching.[570]

In our laboratory, the influence of somatic development on the basic motor skills in boys and girls was studied.[457] In total, 294 6-year-old children from Tartu were studied. Their anthropometric measurements were taken according to the O-Scale physique assessment system.[665] In total, 8 skinfolds, 10 girths, and 2 breadths were measured. Sprint running, standing broad jump, and overarm throwing were recorded using a Panasonic videocamera. Qualitative evaluation of running, throwing, and jumping performance was done by visual observation of videotapes by experienced specialists. Scores for every trial ranged from one through four corresponding to stages one through four.[498,499] Zero-order correlation analysis indicated that there were only a few significant relationships between somatic development and motor skills performances in both boys and girls. Girth parameters mostly influenced the throwing stage. It was concluded that there appear to be significant differences in motor skills between boys and girls in prepubertal years.[457]

One common explanation for the developmental differences between the sexes is that girls do not have the same amount of experience in throwing activities and games as boys.[255,446] Furthermore, several environmental factors such as cultural expectations[580] and rearing factors[173] should be considered when investigating gender differences in motor skills of prepubertal children.

5.4   General considerations

The development of fundamental motor skills among children during pre-pubertal years is an important contributor to successful and satisfying participation in different physical activities later in life. Fundamental motor skills are prerequisites to the advanced skills employed in competitive sports, different games, dance, gymnastics, and other physical activities. Fewer competing activities during prepubertal years give children more time to concentrate on developing motor skills. This period is very important as early detection of motor problems and start of intervention programs can eliminate or minimize many physical and related emotional problems. However, investigations to date have mostly been cross-sectional, which cannot evaluate the developmental process of motor skills. More longitudinal investigations are needed to follow changes in motor skill development during prepubertal years to better understand the various reasons underpinning these changes.

Further information concerning the relationships between motor skills and physical activity, motor ability, and somatic development during prepubertal years is needed. Research to date has shown that several environmental factors influence the acquisition of motor skills, but the specificity of physical activity and instructions should also be considered. While the influence of somatic characteristics on motor skills development seems to be relatively low, more longitudinal investigations on the relationships between motor skills with specific biological and environmental factors are needed.

chapter six

Conclusions and perspectives

Human life and motion cannot be viewed separately. Children are born to move and are spontaneously active in early life. Health and development of children at early ages depend on physical activity and movement possibilities. This important understanding is characterized by the fact that more and more research is focused on studying different aspects of children's movement within the scientific world. Comprehensive publications such as *Pediatric Exercise Science and Medicine* (N. Armstrong and W. Van Mechelen, Eds.) and *Childhood Obesity, Prevention and Treatment* (by J. Parizkova and A. P. Hills) were published in 2000.[19,468] *Body Composition Assessment in Children and Adolescents* (T. Jürimäe and A. P. Hills, Eds.),[317] was released in early 2001.

Recently, the U.S. Department of Health and Human Services[647] emphasized that young people must be taught the skills, knowledge, attitudes, and behaviors that lead to regular participation in physical activity. It is very important to improve elementary motor skills such as correct running and walking, and to learn more difficult skills such as different sports games, cross-country skiing, swimming, etc., during prepubertal years. Emphasis should also be placed on practicing a variety of enjoyable activities. It is especially important that habitually sedentary children find the prescribed physical activities to have fun together with friends and, thereby, become more active.

Prepubertal children need good handbooks to increase their knowledge about the influence of physical activity on their health. They need theoretical knowledge of how to run, jump, throw, play, swim, etc., to avoid elementary mistakes in these skills. There should be chapters with elementary rules of how to dress properly, exercise safely, test different motor abilities individually, and calculate the level of physical activity. Children should also know the history of famous olympic winners of the world as well as their own country. All of this knowledge is important in increasing their willingness and ability to participate in physical activity, to form good health habits, and to get acquainted with various forms of activity for leisure time. Children's interest in physical activity increases when they visit different sports competitions with parents or with peers.

Physical activity is certainly a health-related behavior in humans. A cumulative 30 minutes of moderately intense physical activity every day, which is recommended for adults, is not applicable to prepubertal children since moderate physical activity is not common for children of this age group. They like short bursts of more vigorous or high-intensity physical activities. These bouts of vigorous physical activity may last less than 15 seconds. In contrast, prepubertal children do not remain inactive for extended periods of time. This demonstrates the highly transitory nature of children's physical activity and is probably necessary for normal growth and development. Accordingly, it is recommended that healthy prepubertal children accumulate by the end of day as much moderate to vigorous physical activity as possible.

School physical education lessons have the greatest potential for reaching the largest number of prepubertal children with organized physical activity programs. Children should spend nearly every minute during physical education lessons in enjoyable moderate to vigorous physical activity and during different sports games. Sports games also stimulate teamwork in prepubertal children. Physical education lessons should be scheduled almost every day in a school timetable to achieve the necessary knowledge and level of activities. It is also recommended that children study new skills via movement and games rather than via long explanations by a teacher. The basic aim of physical education lessons in school should be the promotion of a lifelong habit of aerobic exercise. School curricula should not overemphasize physical activities that selectively eliminate children who are less skilled. The efficiency of physical education lessons greatly depends on the qualifications of a teacher. As a rule, classroom teachers who teach physical education in early grades need more knowledge. The influence of school leaders to create a positive atmosphere of physical education lessons is also high. Aside from increasing the level of physical activity in children, the school is also a place for selection of sports talents. The selection of talented children and recommending them to the sports schools and clubs is important in prepubertal years.

The safety of physical activity programs is of less concern in prepubertal children than in adults. As a rule, children generally halt physical activity before they suffer a dangerous level of fatigue. Minimizing injury risk during physical activity is one of the main tasks of physical educators. Safety is paramount. However, it must be taken into account that most of the injuries in children are minor. A prudent approach will minimize overuse injuries and more serious trauma. Anecdotally, physicians have denied participation in physical education lessons for several weeks for children with, for example, slight finger injuries. These children should be able to participate in selected exercises without any problem. It is essential to increase the knowledge of physicians and parents in understanding the relationships between different health problems and physical activity in children. For example, parents should understand the importance of safety and injury prevention when organized sports are conducted for prepubertal children. Warm-up and stretching exercises to minimize hamstring pulls and similar injuries should become habitual preludes to strenuous exercise. The bones in prepubertal children frequently grow at a faster rate than adjacent muscles and tendons and predispose children to muscle tightness, especially at the hamstrings and quadriceps. In addition, frequent recreation-related injuries to children are caused by motor vehicles. Play areas should be away from traffic and safe practices emphasized for walking and biking. The use of helmets during bicycling and in-line skating is also required.

The promotion of higher levels of physical activity and movement in prepubertal years is justified. However, many parents lose interest in further motor development once their child is able to walk unassisted. They may think that with further growth the child will automatically learn new skills, all necessary motor abilities will develop without any assistance, and spontaneous physical activity is enough for normal growth and development. However, children need assistance and encouragement from their parents in learning new skills and developing different motor abilities. Well-developed motor abilities predispose children to more intensive involvement in different physical activities, which become more enjoyable with less strain. This also encourages children to remain physically active throughout life.

The general assumption that more active individuals are more fit applies also to prepubertal children. However, it is necessary to distinguish between the terms of physical activity and physical fitness. This is especially important in smaller children as physical activity is a behavior, whereas physical fitness is an attribute. Physical fitness is also affected by genetic inheritance and maturational status in addition to the level of physical activity. Children should be exposed to the principles of routine exercise and physical activity as early as possible in preschool, school, and family settings. Outdoor physical activities should also be emphasized. Children have achieved most of the elementary motor skills, and they like to exercise and are relatively physically active during prepubertal years. However, some elements of spontaneous physical activity that are unstructured (running during the breaks at school) still remain during prepubertal years. Most children like to participate in different competitions. However, these activities must emphasize self-improvement, participation, and cooperation instead of winning and losing.

Pediatric specialists need more information about the level of basic motor abilities with different age and sex groups of prepubertal children. It must be taken into account that prepubertal children differ from adults in their physical growth, cognitive ability, and psychological status. In addition, 9- to 11-year-old children differ from preschool children. These children already understand what they have to do during testing, and it is possible to motivate them more to exercise with maximal effort. However, prepubertal children need a longer familiarization period before testing in comparison with older children and adults. Differences in most motor performance tests between prepubertal children and older individuals result from biomechanical rather than physiological factors.

Testing is important as a part of comprehensive fitness curricula to teach prepubertal children about the health-related benefits of exercise. However, the selection of different motor ability tests for health-related physical fitness is rather complicated for prepubertal children. Universally accepted fitness tests that are suitable for use in a school environment and that provide objective measures of fitness are simply not available. Test batteries have been presented in different countries that contain varying tests for measuring basic motor abilities (endurance, strength, speed, etc.). A simple and rapid test battery is needed for measuring health-related motor abilities. Furthermore, we need highly standardized and scientifically accepted tests for comparison of motor development in children among different countries and/or different parts of countries. Differences are relatively large in motor ability parameters depending on geographical location and socioeconomic situation. However, it would be incorrect for physical education teachers to overemphasize the results of different motor ability tests. These results are important in helping to improve the motor abilities in prepubertal children who are not fit.

It is clear that peak athletic performance in most sports events can only be obtained by systematic training beginning during prepubertal years. Some children begin to exercise intensively at the ages of 4 or 5 years in some sports events. However, in general, sports specialization should be avoided in prepubertal years. Universally accepted motor ability test batteries for sports talent selection in prepubertal children are not yet available. It has been suggested that final talent identification for most sports events should not take place before puberty. In contrast, organized sports are important for prepubertal children because sport competitions can play an important role in socialization, self-esteem, and self-perception. Sports also establish the basis for a healthy lifestyle and lifelong commitment to physical activity.

Effective and relatively simple methods for the measurement of body composition are lacking for prepubertal children since they are chemically immature. Prior to sexual maturation, children have more water and less bone mineral content than adults, and the density of the fat-free mass changes from prepubertal years to adulthood. However, some body composition methods and regression equations assume that the individuals being measured differ from each other only in the amount of body fat, while the density of fat-free mass is the same for all individuals. This makes it important to carefully consider the assessment technique of body composition in prepubertal children. The simple calculation of BMI is relatively acceptable to assess the somatic growth of prepubertal children. However, the use of skinfold thickness measurements requires preliminary preparation and experience. At present, the bioelectrical impedance analysis procedure seems most appropriate for the assessment of body composition in prepubertal children when using age- and sex-specific regression equations. Body composition assessment is especially important in determining the possible extent of excess weight in prepubertal years.

The assessment of excess weight in children is important in the early diagnosis and prevention of conditions that are associated in adulthood with different health problems. Overweight children are at increased risk of many health problems, including hypertension, hyperlipidemia, and diabetes. Obesity definitions have not been clearly established for prepubertal children. It is suggested that subcutaneous fat is very unstable during infancy and early childhood. However, the fattest children after 6 years of age have a higher risk of remaining fat through childhood and into adulthood. Accordingly, the risk of excess fat appears to be greater for those who have thicker subcutaneous fat measurements during childhood. The relationships between physical activity and adiposity in children are complex, especially at earlier ages. Increasing the level of physical activity while restricting caloric intake has been documented as an effective weight loss strategy. Obese children are frequently less active than children with normal body mass.

Relationships are significant among physical activity, normal growth, and motor development in prepubertal children. The amount of physical activity is moderately but significantly related to aerobic fitness, although physical activity patterns are often characterized by short-burst, predominantly anaerobic activities. Data is lacking about the relationships between physical activity and other motor ability parameters in prepubertal children. However, the results of motor ability tests that need speed and strength are lower in children with smaller stature, while the results of motor ability tests that need endurance are relatively low in children who are tall and present high body mass values. Different somatic characteristics have been shown to also influence motor skill development. For example, overhand throwing development is reportedly related to body stature. In contrast, it has been argued that several sociocultural factors and specific physical activities related to the performance of skill-specific movements have more impact on the qualitative form of motor skill patterns in prepubertal children.

The different aspects of normal growth and development in prepubertal children must be considered. A continuum of main parameters of optimal health and development in prepubertal children is presented in Figure 6.1. The evaluation criteria of normal growth and development in prepubertal children should consider all these parameters. At present, the selection of parameters to assess the optimal health and development in prepubertal ages used in our investigations is only a modest addition that should be developed further. In the future, more comprehensive investigations, including all aspects of optimal health and development, should be conducted to follow the normal growth in prepubertal years longitudinally.

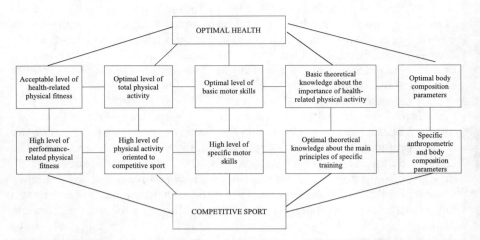

Figure 6.1 A continuum of main parameters of optimal health and successive participation in sport competitions in prepubertal childen.

Recommendations for future research

1. Longitudinal investigations that include different test batteries to measure biological maturation, body composition, physical activity, motor ability and motor skill parameters in prepubertal years
2. Better selection of physical activities that children prefer and then the selection of some of these physical activities for exercise prescriptions
3. Development of test batteries that are simple, not time consuming, and understandable for children when measuring health-related motor abilities
4. Better criteria for sports talent selection
5. Better recommendations on how to effectively teach elementary and/or more complicated motor skills in children
6. Highly valid measures of physical activity; monitors to enable measurement of upper body activities (throwing, catching, etc.); the need for a golden standard is imperative
7. The maximal training loads that do not negatively influence the health of children in sports events such as gymnastics
8. Teacher's or instructor's educational level (qualification) to improve the efficiency of compulsory physical education lessons
9. Parents' educational level to improve their involvement in the complex development of children

Appendices

Appendix 1   Recommended Standard Kits for the Measurement of Anthropometric Parameters in Children[a]

Ross[528]	Norton et al.[448]	Ward et al.[665]	Lohman et al.[381]
Skinfolds			
Triceps	Triceps	Triceps	Triceps
Biceps	Subscapular	Subscapular	Subscapular
Subscapular	Biceps	Biceps	Supraíliac
Iliac crest	Iliac crest	Iliac crest	Abdominal
Supraspinale	Supraspinale	Supraspinale	
Abdominal	Abdominal	Abdominal	
Front thigh	Front thigh	Front thigh	
Medial calf	Medial calf	Medial calf	
	Mid axilla		
Girths			
Arm relaxed	Head	Arm relaxed	Chest
Arm flexed	Neck	Arm flexed and tensed	Hip
Forearm	Arm relaxed	Forearm (max. relaxed)	Arm
Wrist	Arm flexed and tensed	Wrist (distal styloid)	
Head	Forearm	Chest (mesosternale)	
Neck	Wrist	Waist (min.)	
Chest	Chest	Gluteal (max.)	
Waist	Waist	Thigh (1 cm dist. glut. line)	
Omphalion	Gluteal	Calf (max.)	
Hip	Thigh	Ankle	
Thigh	Thigh mid-troch-tibiale-laterale		
Mid-thigh	Calf		
Calf	Ankle		
Ankle			
Lengths			
Arm (acr-rad)	Acromiale radiale		Upper extremity
Forearm (rad-sty)	Radiale-stylion		
Hand (msty-dac)	Midstylion-dactylion		
Isp-box	Iliospinale box height		
Tro-box	Trochanterion box height		
Thigh (tro-til)	Trochanterion tibiale-laterale		
Leg (til-box)	Tibiale-laterale to floor		
Tibia (tim-sphm)	Tibiale mediale-sphyrion-tibiale		
Foot (ak-pt)			

Appendix 1 Continued

Ross [528]	Norton et al. [448]	Ward et al. [665]	Lohman et al. [381]
		Breadths/Lengths	
Biacromial	Biacromial	Humerus	Biacromial
Biiliocristal	Biiliocristal	Femur	Biiliac
Trans chest	Foot length		
AP-chest	Sitting height		
Humerus	Transverse chest		
Wrist	A-P chest depth		
Hand	Humerus		
Femur	Femur		

[a]For more detailed information about the anthropometric landmarks, see original publications. For full anthropometric profiles, use the recommendations of the International Society for Advancement Kinanthropometry.[448]

Appendix 2 Recommended Methods and Prediction Equations for the Measurement of Body Composition in Prepubertal Children

Method	Gender	Age	Prediction Equations	Reference
Skinfolds Σtriceps + calf SF[c]	M[a] F[d]	6 to 17 6 to 17	$\%BF^b = 0.735\,(\Sigma SF) + 1.0^{a,c}$ $\%BF^b = 0.610\,(\Sigma SF) + 5.1^{a,c}$	Slaughter et al.[597] Slaughter et al.[597]
BIA[e] BIA[e]	M/F[a,d] M/F[a,d]	7 to 9 10 to 19	$FFM = 0.640 \times 10^4 \times S^2/R + 4.83^{f,g,h}$ $FFM = 0.61\,(S^2/R) + 0.25\,(BM) + 1.31^{f,g,h,i}$	Deurenberg et al.[159] Houtkooper et al.[289]
BIA[e], SF[c] and anthropometry	M[a]	7 to 25	$FFM = -2.9316 + 0.6462\,(BM) - 0.1159\,(\text{calf SF}) - 0.3753\,(\text{midaxillary SF}) + 0.4754\,(\text{arm circumference}) + 0.1563\,(S^2/R)^{c,f,g,h,i}$	Guo et al.[248]
	F[d]	7 to 25	$4.3383 + 0.6819\,(BM) - 0.1846\,(\text{calf SF}) - 0.2436\,(\text{triceps SF}) - 0.2018\,(\text{subscap. SF}) + 0.1822\,(S^2/R)^{c,g,h,i}$	Guo et al.[248]

[a] M — male.
[b] %BF — percent body fat.
[c] SF — skinfold.
[d] F — female.
[e] BIA — bioelectrical impedance analysis.
[f] FFM — fat free mass.
[g] S — stature.
[h] R — resistance.
[i] BM — body mass.

Appendix 3   A Suggested Combination of Methods (in order of importance) to Measure Daily Physical Activity in Prepubertal Children

Number of Subjects	Measurement Method
<20	1) direct observation 2) doubly labeled water 3) indirect calorimetry
20-100	1) motion sensors 2) heart rate monitors 3) questionnaires (using the help of parents and teachers)
>100	1) questionnaires (using the help of parents and teachers) 2) motion sensors

Source: Modified from Saris, W. H. M., *Acta Paediatr. Scand.,* 318, 37, 1985.

Appendix 4 Recommended Sample Test Battery for the Measurement of Health-Related Physical Fitness in Prepubertal Children

Measure	Influence	Recommended Tests
Cardiorespiratory endurance	lower risk for coronary heart disease (CHD), hypertension, type 2 diabetes mellitus and other chronic degenerative diseases	1-mile run or 20-meter endurance shuttle run
Muscular strength and endurance	reduced fatigue reduced risk for musculoskeletal injury	sit-ups during 1 minute pull-up or push-up
Flexibility	reduced risk for low back pain other musculoskeletal problems	sit-and-reach
Body composition	linked with a lower risk of number of chronic degenerative diseases (CHD, cancer, type 2 diabetes mellitus, hypertension)	sum of subscapular and triceps skinfolds or sum of calf and triceps skinfolds

References

1. Abernethy, B., Kippers, V., Mackinnon, L. T., Neal, R. J., and Hanrahan, S., *The Biophysical Foundations of Human Movement*, Human Kinetics, Champaign, 1997.
2. Adams, F. H., Factors affecting the working capacity of children and adolescents, in *Physical Activity: Growth and Development*, Rarick, A. L., Ed., Academic Press, New York, 1973, 81.
3. Ainsworth, B. E., Haskell, W. H., Leon, A. S., Jacobs, J. R., Montoye, H. J., Sallis, J. F., and Paffenbarger, J. R., Compendium of physical activities: classification of energy costs of human physical activities, *Med. Sci. Sports Exercise*, 25, 71, 1994.
4. Akgün, N., Eurofit test results in the western part of Turkey, in *The Eurofit Tests of Physical Fitness*, Ismir, 1990, 69.
5. Al-Hazzaa, H. M., and Sulaiman, M. A., Maximal oxygen uptake and daily physical activity in 7- to 12-year-old boys, *Pediatr. Exercise Sci.*, 5, 357, 1993.
6. American Alliance for Health, Physical Education, Recreation and Dance, *Technical Manual: Health Related Fitness*, Reston, 1984.
7. American Alliance for Health, Physical Education, Recreation and Dance, *Physical Best: A Physical Fitness Education & Assessment Program*, Reston, 1988.
8. American Alliance for Health, Physical Education, Recreation and Dance, *Physical Best: The AAHPERD Quide to Physical Fitness Education and Assessment*, Reston, 1989.
9. American College of Sports Medicine, Opinion statement on physical fitness in children and youth, *Med. Sci. Sports Exercise*, 20, 422, 1988.
10. American College of Sports Medicine, *Guidelines for Exercise Testing and Prescription*, Lea & Febiger, Philadelphia, 1991.
11. American Health and Fitness Foundation, *Fit Youth Today*, Austin, 1986.
12. Ames, L. B., The sequential patterning of prone progression in the human infant, *Genetic Psychol. Monogr.*, 19, 409, 1937.
13. Armstrong, N., Physical education and health with reference to growth and maturation, in *Active Living Through Quality Physical Education*, Fisher, R., Laws, C., and Moses, J., Eds., proceedings of the Eighth European Congress of ICHPER, London, 16, 1998.
14. Armstrong, N., Balding, J., Gentle, P., and Kirby, B., Patterns of physical activity among 11- to 16-year-old British children, *Br. Med. J.*, 301, 203, 1990.
15. Armstrong, N., Balding, J., Gentle, P., Williams, J., and Kirby, B., Peak oxygen uptake and physical activity in 11 to 16 year olds, *Pediatr. Exercise Sci.*, 2, 349, 1990.

16. Armstrong, N. and Bray, S., Primary schoolchildren's physical activity patterns during autumn and summer, *Bull. Phys. Educ.*, 26, 23, 1990.

17. Armstrong, N., Welsman, J. R., Nevill, A. M., and Kirby, B. J., Health-related physical activity in the national curriculum., *Br. J. Phys. Educ.*, 21, 225, 1990.

18. Armstrong, N., Welsman, J. R., Nevill, A. M., and Kirby, B. J., Modelling growth and maturation changes in peak oxygen uptake in 11- to 13-year-olds, *J. Appl. Physiol.*, 87, 2230, 1999.

19. Armstrong, N. and Van Mechelen, W., *Pediatric Exercise Science and Medicine*, Oxford University Press, New York, 2000.

20. Arnot, R. and Gaines, C., *Sport Talent*, Penguin Books, New York, 1986.

21. Atomi, Y., Iwaoka, K., Hatta, H., Miyashita, M., and Yamamoto, Y., Daily physical activity levels in preadolescent boys related to VO_{2MAX} and lactate threshold, *Eur. J. Appl. Physiol.*, 55, 156, 1986.

22. Bailey, D. A., Exercise, fitness and physical education for the growing child: a concern, *Can. J. Public Health*, 64, 421, 1973.

23. Bailey, D. A., and Mirwald, E. L., A children's test of fitness, *Med. Sport*, 11, 56, 1978.

24. Bailey, R. C., Olson, J., Pepper, S. L., Porszasz, J., Barstow, T. J., and Cooper, D. M., The level and tempo of children's physical activities: an observational study, *Med. Sci. Sports Exercise*, 27, 1033, 1995.

25. Bajin, B., Talent identification program for Canadian female gymnasts, in *World Identification for Gymnastic Talent*, Petiot, B., Salmela, J. H., and Hoshizaki, T. B., Eds., Sport Psyche Editions, Montreal, 1987, 34.

26. Bandini, Z. G., Vu, D. M., Must, A., and Dietz, V. H., Body fatness and bio-electrical impedance in non-obese pre-menarcheal girls: comparison to anthropometry and evaluation of predictive equations, *Eur. J. Clin. Nutr.*, 51, 673, 1997.

27. Barabas, A., Schoolchildren's physical performance and socioeconomic relationships, in Physical Education and Life-long Physical Activity, Proc. of Jyväskylä Sport Congress "Movement and Sport," Report on Physical Culture and Health, Telama, R., Laakso, L., Pieron, M., Ruoppila, I., and Vihko, V., Eds., Jyväskylä, 422, 1990.

28. Barabas, A. and Eiben, O. G., Physical performance abilities of the Hungarian schoolchildren with special regard to the relative strength, in *Methods of Functional Anthropology* (2), Novotony, V. V. and Titlbachova, S., Eds., University Carolina, Pragensis, Prague, 207, 1990.

29. Barabas, A. and Eiben, O. G., Changes in physical performance related to age and biological development, in *Children and Exercise, Pediatric Work Physiology XV*, Frenkl, R. and Szmodis, I., Eds., National Institute for Health Promotion, Budapest, 100, 1991.

30. Baranowski, T., Validity and reliability of self-report measures of physical activity: an information-processing perspective, *Res. Q. Exercise Sport*, 59, 314, 1988.

31. Baranowski, T., Bouchard, C., Bar-Or, O., Bricker, T., Heath, G., Kimm, S. Y. S., Malina, R., Obarzanek, E., Pate, R., Strong, W. B., Truman, B., and Washington, R., Assessment, prevalence, and cardiovascular benefits of physical activity and fitness in youth, *Med. Sci. Sports Exercise*, 24, S237, 1992.

32. Barillas-Mury, C., Vettorazzi, C., Molina, S., and Pineda, O., Experience with bioelectrical impedance analysis in young children: sources of variability, in *In Vivo Body Composition Studies,* Ellis, K. J., Yasumura, S., and Morgan, W. D., Eds., Brookhaven Institute, Upton, New York, 87, 1987.

33. Bar-Or, O., Pathophysiologic factors which limit the exercise capacity of the sick child, *Med. Sci. Sports Exercise,* 18, 276, 1986.

34. Bar-Or, O., A commentary to children and fitness — a public health perspective, *Res. Q. Exercise Sport,* 58, 304, 1987.

35. Bar-Or, O., Physical activity and physical training in childhood obesity, *J. Sports Med. Phys. Fitness,* 33, 323, 1993.

36. Bar-Or, O. and Baranowski, T., Physical activity, adiposity, and obesity among adolescents, *Pediatr. Exercise Sci.,* 6, 346, 1994.

37. Batterham, A. M., Tolfrey, K., and George, K. P., Nevill's explanation of Kleiber's 0.75 mass exponent: an artifact of colinearity problems in least square model?, *J. Appl. Physiol.,* 82, 693, 1997.

38. Baumgartner, R. N., Chumlea, W. C., and Roche, A. F., Associations between bioelectrical impedance and anthropometric variables., *Hum. Biol.,* 59, 235, 1978.

39. Baumgartner, R. N. and Jackson, A., *Measurement for Evaluation in Physical Education and Exercise Science,* Brown, Dubuque, 1987.

40. Baumgartner, R. N., Chumlea, W. C., and Roche, A. F., Bioelectric impedance phase angle and body composition, *Am. J. Clin. Nutr.,* 48, 16, 1988.

41. Baumgartner, R. N., Chumlea, W. C., and Roche, A. F., Estimation of body composition from segment impedance, *Am. J. Clin. Nutr.,* 50, 221, 1989.

42. Baumgartner, R. N., Chumlea, W. C., and Roche, A. F., Bioelectric impedance for body composition, *Exercise Sci. Rev.,* 18, 193, 1990.

43. Baumgartner, R. N., and Roche, A. F., Tracking of fat pattern indices in childhood: the Melbourne growth study, *Hum. Biol.,* 60, 549, 1998.

44. Baxter-Jones, A. D. G., Growth and development of young athletes: should competition levels be age related?, *Sports Med.,* 20, 59, 1995.

45. Bayley, N., *The California Infant Scale of Motor Development,* University of California Press, Berkeley, 1936.

46. Bayley, N., The accurate prediction of growth and adult height, *Mod. Probl. Pediatr.,* 7, 224, 1962.

47. Bell, W., Body size and shape: a longitudinal investigation of active and sedentary boys during adolescence, *J. Sports Sci.,* 11, 127, 1993.

48. Berenson, G. S., McMahan, C. A., and Voors, A. V., *Cardiovascular Risk Factors in Children: the Bogalusa Heart Study,* Oxford University Press, New York, 1980.

49. Bettiol, H., Rona, R. J., and Chinn, S., Variation in physical fitness between ethnic groups in nine year olds, *Int. J. Epidemiol.,* 28, 281, 1999.

50. Beunen, G., Biological age in pediatric exercise research, in *Advances in Pediatric Sport Sciences,* Vol. 3, *Biological Issues,* Bar-Or, O., Ed., Human Kinetics, Champaign, 1989.

51. Beunen, G. Ostyn, M., Simons, J., Van Gerven, D., Swalus, P., and De Beul, K., A correlational analysis of skeletal maturity, anthropometric measures and motor fitness of boys 12 through 16, in *Biomechanics of Sport and Kinanthropometry,* Landry, F. and Orban, W. A. R., Eds., Symposia Specialists, Miami, 1978, 343.

52. Beunen, G., Ostyn, M., Simons, J., Renson, R., and Van Gerven, D., Cronological age and biological age as related to physical fitness in boys 12 to 19 years, *Ann. Hum. Biol.,* 8, 321, 1981.

53. Beunen, G. and Claessens, A., Physical Fitness Evaluatie: De PF-Leuven Test Batterij [Physical Fitness Evaluation. The PF-Leuven Test Battery], *Gen. Sport,* 6, 224, 1987.

54. Beunen, G. and Malina, R.M., Growth and physical performance relative to the timing of the adolescent spurt, *Exercise Sport Sci. Rev.,* 16, 503, 1988.

55. Beunen, G., Lefevre, A., and Claessens, A., Age-specific correlation analysis of longitudinal physical fitness levels in men, *Eur. J. Appl. Physiol.,* 64, 538, 1992.

56. Beunen, G., Malina, R. M., Renson, R., Simons, J., Ostyn, M., and Lefevre, J., Physical activity and growth, maturation and performance: a longitudinal study, *Med. Sci. Sports Exercise,* 24, 576, 1992.

57. Beunen, G. and Malina, R. M., Growth and biological maturation: relevance to athletic performance, in *The Encyclopedia of Sports Medicine: The Child and Adolescent Athlete,* Bar-Or, O., Ed., Blackwell Scientific Publications, Oxford, 1996, 3.

58. Beunen, G. Malina, R. M., Lefevre, J., Claessens, A. L., Renson, R., Kanden Eynde, B., Vanreusel, B., and Simons, J., Skeletal maturation, somatic growth and physical fitness in girls 6 to 16 years of age, *Int. J. Sports Med.,* 18, 413, 1997.

59. Beunen, G., Rogers, D. M., Wynarowska, B., and Malina, R. M., Longitudinal study of ontogenetic allometry of oxygen uptake in boys and girls grouped by maturity status, *Ann. Hum. Biol.,* 24, 33, 1997.

60. Biddle, S. J. H. and Fox, K. R., The child's perspective in physical education. Part 4: achievement psychology. *Br. J. Phys. Educ.,* 19, 182, 1988.

61. Bielicki, T., Koniarek, J., and Malina, R. M., Interrelationships among certain measures of growth and maturation rate in boys during adolescence, *Ann. Hum. Biol.,* 11, 201, 1984.

62. Björntorp, P., Abdominal fat distribution and disease: an overview of epidemiological data, *Ann. Med.,* 24, 15, 1992.

63. Blaha, P., Šrajer, J., and Krasnicanova, H., Czech obese children — loss of body weight during reducing treatment, Papers on Anthropology VII, Tartu University Press, Tartu, 1997, 64.

64. Blair, S. N., Are American children and youth unfit? The need for better data. *Res. Q. Exercise Sport,* 63, 120, 1992.

65. Blair, S. N., Physical activity interventions with children and youth, in *Prevention of Atherosclerosis and Hypertension Beginning in Youth,* Filer, L. J., Lauer, R. M., and Luepker, R. V., Eds., Lea & Febiger, Philadelphia, 1994, 273.

66. Blair, S. N., Clark, D. G., Cureton, K. J., and Powell, K. E., Exercise and fitness in childhood: implications for a lifetime of health, in *Perspectives in Exercise Science and Sports Medicine,* Gisolfi, C. V., and Lamb, D. R., Eds., Benchmark, Indianapolis, 1989, 401.

67. Blair, S. N., Kohl, H. W., Paffenbarger, R. S. Jr., Clark, D. G., Cooper, K. H., and Gibbons, L. W., Physical fatness and all-cause mortality, a prospective study of healthy men and women, *JAMA,* 262, 2395, 1989.

68. Blair, S. N. and Meredith, M. D., The exercise-health relationship: does it apply to children and youth? in *Health and Fitness Through Physical Education,* Pate, R. R. and Hahn, R. C., Eds., Human Kinetics, Champaign, 1994.

69. Blanksby, B. A., Bloomfield, J., Ackland, T. R., Elliott, B. C., and Morton, A.R., *Athletics, Growth, and Development in Children, The University of Western Australia Study,* Harwood Academic Publishers, 1993.

70. Blomquist, B., Borjeson, M., Larsson, Y., Persson, B., and Sterky, G., Effect of physical activity on the body measurements and work capacity of over-weight boys, *Acta Paediatr. Scand.,* 54, 566, 1965.

71. Bloom, B. S., *Stability and Change in Human Characteristics,* John Wiley & Sons, New York, 1964.

72. Bodzsar, E. B., The indices of the physique and the socio-economic factors based on growth in Bakony girls, *Anthropol. Közl.,* 26, 129, 1982.

73. Bodzsar, E. B., Secular growth changes in Hungary, in *Secular Growth Changes in Europe,* Bodzsar, E. B. and Susanne, C., Eds., Eötvös University Press, Budapest, 175, 1998.

74. Boileau, R. A., Utilization of total body electrical conductivity in determining body composition, in *National Research Council, Designing Foods: Animal Product Options in the Marketplace,* National Academy, Washington, 1988, 251.

75. Boileau, R. A., Lohman, T. G., Slaughter, M. H., Ball, T. E., Going, S. B., and Hendrix, M. K., Hydration of the fat-free body in children during matura-tion, *Hum. Biol.,* 56, 651, 1984.

76. Boileau, R. A., Wilmore, J. H., Lohman, T. G., Slaughter, M. H., and Riner, W. F., Estimation of body density from skinfold thicknesses, body circumferences and skeletal widths in boys aged 8 to 11 years, comparison of two samples, *Hum. Biol.,* 53, 575, 1981.

77. Bompa, T. O., Talent identification, in *Sports: Science Periodical on Research and Technology in Sport, Physical Testing G1,* Coaching Association of Canada, Ottawa, 1985.

78. Boot, A. M., Bouquet, J., de Ridder, M. A. J., Krenning, E. P., and de Muinck Keizer-Schrama, S. M. P. F., Determinants of body composition measured by dual energy x-ray absorptiometry in Dutch children and adolescents, *Am. J. Clin. Nutr.,* 66, 232, 1997.

79. Booth, M. L., Okely, T., McLellan, L., Phongsavan, P., Macaskill, P., Patterson, J., Wright, J., and Holland, B., Mastery of fundamental motor skills among New South Wales school students: prevalence and sociodemographic distri-bution, *J. Sci. Med. Sport,* 2, 93, 1999.

80. Boreham, C., Savage, M., Primrose, D., Cran, G., and Strain, J., Coronary risk factors in schoolchildren, *Arch. Dis. Child.,* 68, 182, 1993.

81. Boreham, C. A., Twisk, J., Savage, M. J., Cran, G. W., and Strain, J. J., Physical activity, sports participation, and risk factors in adolescents, *Med. Sci. Sports Exercise,* 29, 788, 1997.

82. Bouchard, C., Genetics of aerobic power and capacity, in *Sport and Human Genetics,* Malina, R. M. and Bouchard, C., Eds., Human Kinetics, Champaign, 1986, 59.

83. Bouchard, C., Leblanc, C., Malina, R. M., and Hollmann, W., Skeletal age and submaximal working capacity in boys, *Ann. Hum. Biol.,* 5, 75, 1978.

84. Bouchard, C. and Malina, R. M., Genetics of physiological fitness and motor performance, *Exercise Sport Sci. Rev.,* 11, 306, 1983.

85. Bouchard, C. and Shephard, R. J., Physical activity, fitness, and health: the model and key concepts, in *Physical Activity, Fitness, and Health International Proceedings and Consensus Statement,* Bouchard, C., Shephard, R. J., Stephens, T., Eds., Human Kinetics, Champaign, 1994, 77.

86. Bouchard, C., Shephard, R. J., and Stephens, T., *Physical Activity, Fitness, and Health: Statement*, Human Kinetics, Champaign, 1994.

87. Bouckaert, J., Van Vytvanck, P., and Vrijiens, J., Anthropometrical data, muscle strength, physiological and selected motor ability factors of 11-year-old boys. *Acta Paediatr. Belgica*, 28, S60, 1974.

88. Bovendeerdt, J., Physical education and health education in the Dutch school-practice (12 to 18 years old), in *Proc. of the 6th ICHPER-Europe Congress*, Prague, 56, 1992.

89. Brand, K., Eggert, D., Jendritzki, H., and Küpper, B., Untersuchungen zur motorischen Entwicklung von Kinder im Grundschulalter in den Jahren 1985 und 1995, *Praxis der Psychomotorik*, 22, 101, 1997.

90. Branta, C., Haubenstricker, J., and Seefeldt, V., Age changes in motor skills during childhood and adolescence, *Exercise Sport Sci. Rev.*, 12, 467, 1984.

91. Brewer, J. P., Balsom, R., and Davis, J., Seasonal birth distribution amongst European soccer players, *Sports Exercise Injury*, 1, 154, 1995.

92. Brio, R., Porpiglia, M., and Chair, A., International medicine, obstetric and gynecological problems related to overweight, *Panminerva Med.*, 36, 138, 1994.

93. Broadhead, G. D. and Church, G. E., Movement characteristics of preschool children, *Res. Q. Exercise Sport*, 56, 208, 1985.

94. Brozek, J., Grande, F., Anderson, J. T., and Kemp, A., Densitometric analysis of body composition: revision of some quantitative assumptions, *Ann. N.Y. Acad. Sci.*, 110, 113, 1963.

95. Brundtland, G. H., Liestol, K., and Walloe, L., Height, weight and menarcheal age of Oslo schoolchildren during the last 60 years, *Ann. Hum. Biol.*, 7, 307, 1980.

96. Buday, J., Growth and physique in Down's syndrome children and adults, *Hum. Biol. Budapest*, 20, 1990.

97. Burton, W. A. and Miller, D. E., *Movement Skill Assessment*. Human Kinetics, Champaign, 1998.

98. Buschner, C. A., *Teaching Children Movement Concepts and Skills: Becoming a Master Teacher*, Human Kinetics, Champaign, 1994.

99. Butcher, J. E. and Eaton, W. O., Gross and fine motor proficiency in preschoolers: relationships with free play behaviour and activity level, *J. Hum. Movement Stud.*, 16, 27, 1989.

100. Butterfield, S. A., and Loovis, E. M., Influence of age, sex, balance, and sport participation on development of throwing by children in grades K-8, *Percept. Motor Skills*, 76, 459, 1993.

101 Calderone, G., Leglise, M., Giampetro, M., and Berlutti, G., Anthropometric measurements, body composition, biological maturation and growth predictions in young female gymnasts of high diagnostic level, *J. Sports Med.*, 26, 263, 1986.

102. Cale, L. and Almond, L., Physical activity levels of young children: a review of the evidence, *Health Educ. J.*, 51, 94, 1992.

103. Cale, L. and Harris, J., Exercise recommendations for children and young people, *Phys. Educ. Rev.*, 16, 89, 1993.

104. Campbell, W. R. and Pohndurf, R. H., Physical fitness of British and United States children, in Larson, K., Ed., *Health and Fitness in Modern World*, Athletic Institute, Chicago, 1961.

105. Canada Fitness Survey, *Canadian Youth and Physical Activity*, Canada Fitness Survey, Ottawa, 1981.

106. *Canadian Standardized Test of Fitness: Operations Manual*, 3rd ed., *Fitness and Amateur Sport*, Government of Canada, Ottowa, 1986.

107. Carron, A. V. and Bailey, D. A., Strength development in boys from 10 through 16 years, *Mon. Res. Child*, 39, 1974.

108. Carter, J. E. L. and Heath, B. H., *Somatotyping — Development and Applications*, Cambridge University Press, London, 1990.

109. Carter, J. E. L., Mirwald, R. L., Heath-Roll, B. H., and Bailey, D. A., Somatotypes of 7- to 16-year-old boys in Saskatchewan, Canada, *Am. J. Hum. Biol.*, 9, 257, 1997.

110. Casey, V. A., Dwyer, J. T., Colman, K. A., and Valadian, I., Body mass index from childhood to middle age: a 50-year follow-up, *Am. J. Clin. Nutr.*, 56, 14, 1992.

111. Caspersen, C. J., Physical activity epidemiology: concepts, methods, and applications to exercise science, *Exercise Sport Sci. Rev.*, 17, 423, 1989.

112. Caspersen, C. J., Powell, K. E., and Christiansen, G. M., Physical activity, exercise and physical fitness: definitions and distinctions for health-related research, *Publ. Health Rep.*, 100, 126, 1985.

113. Cassady, S. L., Nielsen, D. H., Janz, K. F., Wu, Y., Cook, J.S., and Hansen, J. R., Validity of near infrared body composition analysis in children and adolescents, *Med. Sci. Sports Exercise*, 25, 1185, 1993.

114. Chumlea, W. C., Physical growth in adolescence, in *Handbook of Developmental Psychology*, Wollmann, B. B., Stricker, G., Ellman, S. J., Keith-Spiegel, P., and Palermo, D. S., Eds., Prentice Hall, Englewood Cliffs, 1982, 471.

115. Chumlea, W. C., Siervogel, R. M., Roche, A. F., Webb, P., and Rogers, E., Increments across age in body composition of children 10 to 18 years of age, *Hum. Biol.*, 55, 845, 1983.

116. Chumlea, W. C., Baumgartner, R. N., and Roche, A. F., The use of specific resistivity to estimate fat-free mass from segmental body measures of bioelectric impedance, *Am. J. Clin. Nutr.*, 48, 7, 1988.

117. Chumlea, W. C. and Guo, S. S., Bioelectrical impedance and body composition: present status and future directions, *Nutr. Rev.*, 52, 123, 1994.

118. Cilia, G. and Belluca, M., *Eurofit: Test Europei de Attitudine Fisica [Eurofit: European Test of Physical Fitness]*, ISEF, Roma, 1993.

119. Claessens, A., Beunen, G., and Simons, J., Stability of anthroposcopic and anthropometric estimates of physique in Belgian boys followed longitudinally from 13 to 18 years of age, *Ann. Hum. Biol.*, 3, 235, 1986.

120. Clarke, H. H., *Physical and Motor Tests in the Medford Boys' Growth Study*, Prentice Hall, Englewood Cliffs, 1971.

121. Clarke, H. H., Individual differences, their nature, extent, and significance, *Phys. Fitness Res. Dig.*, 3 (4), 1973.

122. Clark, J. E. and Whitall, J., Changing patterns of locomotion: from walking to skipping, in *Development of Posture and Gait across the Lifespan*, Wollacott, M. H., and Shumway-Cook, A., Eds., University of South Carolina Press, Columbia, 1989, 128.

123. Clark, J. E. and Whitall, J., What is motor development? the lessons of history, *Quest*, 41, 183, 1989.

124. Cohen, J., *Statistical Power Analysis for the Behavioural Science*, Academic Press, New York, 1969.
125. Coleman, K. J., Saelenes, B. E., Wiedrich-Smith, M. D., Finn, J. D., and Epstein, L. H., Relationships between TriTrac-R3D vectors, heart rate, and self-report in obese children, *Med. Sci. Sports Exercise*, 29, 1535, 1997.
126. Colley, A., Eglinton, E., and Elliott, E., Sport participation in middle childhood: association with styles of play and parental participation, *Int. J. Sport Psychol.*, 25, 193, 1992.
127. Conger, P. R., Quinney, H. A., Gauthier, R., and Massicotte, D., A comparison of the CAHPER performance test 1966-1980, *CAHPER J.*, 6-11, 12-16 Sept/Oct, 1982.
128. Connolly, K. J., Brown, K., and Basset, E., Developmental changes in some components of a motor skill, *Br. J. Psychol.*, 59, 305, 1968.
129. Cooper Institute for Aerobic Research (CIAR), *The Prudental Fitnessgram Test Administration Manual*, Cooper Institute for Aerobic Research, Dallas, 1992.
130. Corbin, C. B., *A Textbook of Motor Development*, Brown, Dubuque, 1973.
131. Corbin, C. B., Youth fitness, exercise and health: there is much to be done, *Res. Q. Exercise Sport*, 58, 308, 1987.
132. Corbin, C. B., Whitehead, J. R., and Lovejoy, P. Y., Youth physical fitness awards, *Quest*, 40, 200, 1988.
133. Corbin, C. B. and Pangrazi, R. P., *Teaching Strategies for Improving Youth Fitness*, Institute for Aerobic Research, Dallas, 1989.
134. Corbin, C. B. and Pangrazi, R. P., Are American children and youth fit, *Res. Q. Exercise Sport*, 63, 96, 1992.
135. Cordain, L., Whicker, R. E., and Johnson, J. E., Body composition determination in children using bioelectrical impedance, *Growth Dev. Aging*, 52, 37, 1988.
136. Corneish, B. H., Jackobs, A., Thomas, B. J., and Ward, L. C., Optimizing electrode sites for segmental bioimpedance measurements, *Physiol. Meas.*, 20, 241, 1999.
137. Cotton, D. J., An analysis of the NCYFS II modified pull-up test, *Res. Q. Exercise Sport*, 61, 272, 1990.
138. Council for Physical Education for Children. *Physical Activity for Children: a Statement of Guidelines*, NASPE Publications, Reston, 1998.
139. Cumming, G. R., Goulding, D., and Baggley, G., Failure of school physical education to improve cardiovascular fitness, *Can. Med. Assoc. J.*, 101, 69, 1969.
140. Cureton, K. J., Commentary on children and fitness: a public health perspective, *Res. Q. Exercise Sport*, 58, 315, 1987.
141. Cureton, K. J., Boileau, R. A., and Lohman, T. G., Relationship between body composition measures and AAHPER test performances in young boys, *Res. Q. Exercise Sport*, 46, 218, 1975.
142. Cureton, K., Boileau, R., Lohman, T., and Misner, J., Determinants of distance running performance in children: analysis of a path moded, *Res. Q. Exercise Sport*, 48, 270, 1977.
143. Cureton, K. J. and Warren, G. L., Criterion reference standards for youth health-related fitness tests: a tutorial, *Res. Q. Exercise Sport*, 67, 7, 1990.
144. Damon, W. and Hart, D., *Self-understanding in Childhood and Adolescence*, Cambridge University Press, Cambridge, 1988.

145. Danford, L. C., Schoeller, D. A., and Kusher, R. F., Comparison of two bio-electrical impedance analysis models for total body water measurement in children, *Ann. Hum. Biol.*, 19, 603, 1992.

146. Davies, K. L. and Christoffel, K. K., Obesity in preschool and school-age children, *Arch. Pediatr. Adolescent Med.*, 148, 1257, 1994.

147. Davies, K. L., Roberts, T. C., Smith, R. R., Ormond F. III, Pfohl, S. Y., and Bowling, M., *North Carolina Children and Youth Fitness Study, JOPERD*, October, 1994, 65.

148. Davies, P. S. W., Gregory, J., and White, A., Physical activity and body fatness in pre-school children, *Int. J. Obesity*, 19, 6, 1995.

149. Deach, D., Genetic Development of Motor Skills in Children Two through Six Years of Age unpublished manuscript, University of Michigan, Ann Arbor, 1950.

150. Deci, L. and Ryan, R. M., *Intrinsic Motivation and Self Determination in Human Behaviour*, Plenum Press, New York, 1985.

151. Deeheger, M., Rolland-Cachera, M. F., and Fontvieille, A. M., Physical activity and body composition in 10 year old French children: linkage with nutritional intake? *Int. J. Obesity*, 21, 372, 1997.

152. Delozier, M.G., Gutin, B., Wang, I., Basch, C. E., Contento, I., Shea, S., Irioynen, M., Zybert, P., Rips, J., and Pierson, R., Validity of anthropometry and bioimpedance with 4- to 8-year-olds using total body water as the criterion, *Pediatr. Exercise Sci.*, 3, 238, 1991.

153. Dennison, B. A., Straus, J. H., Mellits, E. D., and Charney, E., Childhood physical fitness tests: predictor of adult physical activity levels, *Pediatrics*, 82, 324, 1988.

154. Dezenberg, C. U., Nagy, T. R., Gower, B. A., Johnson, R., and Goran, M. I., Predicting body composition from anthropometry in pre-adolescent children, *Int. J. Obesity*, 23, 253, 1999.

155. Deurenberg, P., International consensus conference on impedance in body composition, *Age Nutr.*, 5, 142, 1994.

156. Deurenberg, P., Multi-frequency impedance as a measure of body water compartments, in *Body Composition Techniques in Health and Disease*, Davies, P. S. W. and Cole, T. J., Eds., Cambridge University Press, Cambridge, 1995, 44.

157. Deurenberg, P., van der Kooy, K., Paling, A., and Withagen, P., Assessment of body composition in 8-11 year old children by bioelectrical impedance, *Eur. J. Clin. Nutr.*, 43, 623, 1989.

158. Deurenberg, P., van der Kooy, K., and Schouter, F. J. M., Body impedance is largely dependent on the intra- and extracellular water distribution, *Eur. J. Clin. Nutr.*, 43, 845, 1989.

159. Deurenberg, P., Kusters, C. S. L., and Smit, H. E., Assessment of body composition by bioelectrical impedance in children and young adults is strongly age-dependent, *Eur. J. Clin. Nutr.*, 44, 261, 1990.

160. Deurenberg, P. and Schutz, Y., Body composition: overview of methods and future directions of research, *Ann. Nutr. Metab.*, 39, 325, 1995.

161. Deurenberg, P., Andreoli, A., and De Lorenzo, A., Multi-frequency bioelectrical impedance: a comparison between the Cole-Cole modeling and Hanai equations, with the classical impedance index approach, *Ann. Hum. Biol.*, 23, 31, 1996.

162. Diaham, B. and Prentice, A., Are modern British children too inactive?, *Br. Med. J.,* 306, 998, 1993.
163. Dietz, W. H. Jr., Childhood obesity: susceptibility, cause, and management, *J. Pediatr.,* 103, 676, 1983.
164. Dietz, W. H. and Gortmaker , S. L., Do we fatten our children at the television set? Obesity and television viewing in children and adolescents, *Pediatrics,* 75, 807, 1995.
165. Di Iorio, B. R. and Terracciano, V., Bioelectrical impedance measurement: errors and artifacts, *J. Ren. Nutr.,* 9, 192, 1999.
166. Dinubile, N. A., Youth fitness-problems and solutions, *Prev. Med.,* 22, 589, 1993.
167. Docherty, D., *Measurement in Pediatric Exercise Science,* Human Kinetics, Champaign, 1996.
168. Docherty, D. and Gaul, C. A., Relationship of body size, physique, and composition to physical performance in young boys and girls, *Int. J. Sports Med.,* 12, 525, 1991.
169. Dohrmann, P., Throwing and kicking ability of 8-year-old boys and girls, *Res. Q. Exercise Sport,* 35, 464, 1964.
170. Dollmann, J., Olds, T., Norton, K., and Stuart, D., The evolution of fitness and fatness in 10-11-year-old Australian schoolchildren: changes in distributional characteristics between 1985 and 1997, *Pediatr. Exercise Sci.,* 11, 108, 1999.
171. Du Randt, R., *Sport Talent Identification and Development and Related Issues in Selected Countries,* University of Port Elizabeth, Port Elizabeth, 1992.
172. Duba, J. L., Toward a developmental theory of children's motivation in sport, *J. Sport Psychol.,* 9, 130, 1987.
173. East, W. B. and Hensley, L. D., The effects of selected sociocultural factors upon the overhand throwing performance of prepubescent children, in *Motor Development,* Clark, J. E., Ed., 1985, 115.
174. Ebbeling, C. J., Hamill, J., Freedson, P. S., and Rowland, T. W., An examination of efficiency during walking in children and adults, *Pediatr. Exercise Sci.,* 4, 36, 1992.
175. Eiben, O. G., Growth and physical fitness of children and youth at the end of the XXth century, preliminary report, *Int. J. Anthropol,* 13, 129, 1998.
176. Eiben, O. G., Barabas, A., and Panto, E., The Hungarian national growth study. I. Reference data on the biological developmental status and physical fitness of 3-18 year-old Hungarian youth in the 1980s, *Hum. Biol. Budapest,* 21, 123, 1991.
177. Elia, M., Body composition analysis: an evaluation of 2 component models, multicomponent models and bedside techniques, *Clin. Nutr.,* 11, 114, 1992.
178. Ellis, J. D., Carron, A. V., and Bailey, D. A., Physical performance in boys from 10 through 16 years, *Hum. Biol.,* 47, 263, 1975.
179. Ellis, K. J., and Shypailo, R. J., Whole body potassium measurements independent of body size, *Basic Life Sci.,* 60, 371, 1993.
180. Ellis, K. J., Shypailo, R. J., Pratt, J. A., and Pond, W. G., Accuracy of dual-energy x-ray absorptiometry for body composition measurements in children, *Am. J. Clin. Nutr.,* 60, 660, 1994.
181. Elsin, R., Siu, M.L., Pineda, O., and Solomons, N. W., Sources of variability in bioelectrical impedance determinations in adults, in *In Vivo Body Composition Studies,* Ellis, K. J., Yasumura, S., and Morgan, W. D., Eds., Brookhaven National Laboratory, Upton, New York, 1987, 184.

182. Engelman, M. E., and Morrow, J. R. Jr., Reliability and skinfold correlates for traditional and modified pull-ups in children grades 3-5, *Res. Q. Exercise Sport*, 62, 88, 1991.

183. Engstrom, L. M. and Fischbein, S., Physical capacity in twins, *Acta Genet. Med. Gemellol.*, 26, 159, 1977.

184. Epstein, L. H., Wing, R. R., Koeske, R., Ossip, D. J., and Beck, S., A comparison of lifestyle change and programmed aerobic exercise on weight and fitness changes in obese children, *Behav. Ther.* 13, 651, 1982.

185. Epstein, L. H., Wing, R. R., Penner, B. C., and Kress, M. J., Effect of diet and controlled exercise on weight loss in obese children, *J. Pediatr,.* 107, 3358, 1985.

186. Epstein, L. H., Valoski, A., Wing, R. R., and McCurly, J., Ten-year follow-up of behavioural family-based treatment for obese children, *JAMA*, 264, 2519, 1990.

187. Epstein, L. H. and Goldfield, G. S., Physical activity in the treatment of childhood overweight and obesity: current evidence and research issues, *Med. Sci. Sports Exercise* 31, S553, 1999.

188. Eriksson, B. O., Gollinck, P. D., and Saltin, B., Muscle metabolism and enzyme activities after training in boys 11-13 years old, *Acta Physiol. Scand.*, 87, 485, 1973.

189. Eriksson, B. O. and Saltin, B., Muscle metabolism during exercise in boys aged 11 to 16 years compared to adults, *Acta Paediatr. Belgica*, 28, S257, 1974.

190. Espenschade, A. S. and Eckert, H. M., *Motor Development*, 2nd ed., Merrill, Sydney, 1980.

191. Eston, R. G., Rowlands, A. V., and Ingledew, D. K., Validation of the Tritrac-R3D^{TM} activity monitor during typical children's activities, in *Children and Exercise: XIX Pediatric Work Physiology*, Vol. 1, Armstrong, N., Kirby, B., and Welsman, J., Eds., E. & F.N. Spon, London, 1997, 132.

192. Eston, R. G., Rowlands, A. V., and Ingledew, D. K., Validity of heart rate, pedometry, and accelerometry for predicting the energy cost of children's activities, *J. Appl. Physiol.*, 84, 362, 1998.

193. Eurofit, *European Tests of Physical Fitness, Council of Europe*, Committee for the Development of Sport, Rome, 1988.

194. Eurofit, *La Bateria Eurofit*, A Catalunya, Barcelona, 1993.

195. Evans, J. and Roberts, G. C., Physical competence and the development of children's peer relations, *Quest*, 39, 23, 1987.

196. Eveleth, P. B. and Tanner, J. M., *Worldwide Variation in Human Growth*, 2nd ed., Cambridge University Press, Cambridge, 1990.

197. Evetovich, T. K., Haush, T. J., Eckerson, J. M., Johnson, G. O., Honsh, T. J., Stont, J. R., Smith, D. B., and Ebersch, K. T., Validity of bioelectrical impedance equations for estimating fat-free mass in young athletes, *J. Strength Cond. Res.*, 11, 155, 1997.

198. Fagard, J., Skill acquisition in children: a historical perspective, in *The Child and Adolescent Athlete*, Bar-Or, O., Ed., Blackwell Science, Oxford, 1996, 74.

199. Fairwheather, H. and Hutt, S. J., On the rate of gain of information in children, *J. Exp. Child. Physiol.*, 26, 216, 1978.

200. Farber, G.B., Body composition in adolescence, in *Human Growth*, Faluner, F. and Tanner, J. M., Eds., Plenum Press, London, 119, 1978.

201. Faulkner, R. A., Maturation, in Docherty, D. Ed., *Measurement in Pediatric Exercise Science*, Human Kinetics, Champaign, 1996, 129.
202. Fellmann, N., Bedu, M., Spielvogel, H., Falgairette, G., Van Praagh, E., Jarrige, J.F., and Coudert, J., Anaerobic metabolism during pubertal development at high altitude, *J. Appl. Physiol.*, 64, 1382, 1988.
203. Fenster, J. R., Freedson, P. S., Washburn, R. A., and Ellison R. C., The relationship between peak oxygen uptake and physical activity in 6- to 8-year-old children, *Pediatr. Exercise Sci.*, 1, 127, 1989.
204. Ferrante, E., Pitzalis, G., Deganello, F., Galastri, E., Sciarpelletti, R., and Imperato, C., The evaluation of body composition in children by anthropometry and impedance measurements, *Minerva Pediatr.*, 45, 289, 1993.
205. Fields, D. A. and Goran, M. I., Body composition techniques and the 4-compartment model in children, *J. Appl. Physiol.*, 89, 613, 2000.
206. Fishbein, S., Onset of puberty in MZ and DZ twins, *Acta Genet. Med. Gemellol.*, 26, 151, 1977.
207. Fitness Canada, Canadian Standardized Test of Fitness: Operations Manual, 3rd ed., *Fitness and Amateur Sport*, Ottawa, 1987.
208. Flint, M. M. and Gudgell, J., Electromyographic study of abdominal muscular activity during exercise, *Res. Q. Exercise Sport*, 36, 29, 1965.
209. Folsom-Meek, S. L., Herauf, J., and Adams, N. A., Relationships among selected attributes and three measures of upper body strength and endurance in elementary school children, *Percept. Motor Skills*, 75, 1115, 1992.
210. Fomon, S. J., Haschke, F., Ziegler, E. E., and Nelson, S. E., Body composition of reference children from birth to age 10 years, *Am. J. Clin. Nutr.*, 35, 1169, 1982.
211. Forbes, G. B., Exercise and body composition, *J. Appl. Physiol.*, 70, 994, 1991.
212. Forbes, G. B., Body composition: influence of nutrition, disease, growth, and aging, in *Modern Nutrition in Health and Disease*, 8th ed., Shils, M. E., Alson, J. A., and Shike, M., Eds., Lea & Febiger, Philadelphia, 1994.
213. Fortney, V. L., The kinematic and kinetics of the running pattern of two, four, and six-year-old children, *Res. Q. Exercise Sport*, 54, 126, 1983.
214. Fox, K. R. and Biddle, S. J. H., Health-related fitness testing in schools: introduction and problems of interpretation, *Bull. Phys. Ed.*, 22, 54, 1986.
215. Fox, K. R. and Biddle, S. J. H., Health-related fitness testing in schools: philosophical and psychological implications, *Bull. Phys. Ed.*, 23, 28, 1987.
216. Fox, K. R. and Biddle, S. J. H., The use of fitness tests: educational and psychological considerations, *JOPERD*, 59, 47, 1988.
217. Frazer, G. E., Phillips, R. L., and Harris, R., Physical fitness and blood pressure in schoolchildren, *Circulation*, 67, 405, 1983.
218. Freedson, P. S., Field monitoring of physical activity in children, *Pediatr. Exercise Sci.*, 1, 8, 1989.
219. Freedson, P. S., Physical activity among children and youth, *Can. J. Sport Sci.*, 17, 280, 1992.
220. Freedson, P. S., and Evenson, S., Familial aggregation in physical activity, *Res. Q. Exercise Sport*, 62, 384, 1991.
221. Freedson, P. S., Cureton, K. J., and Heath, G. W., Status of field-based fitness testing in children and youth, *Prev. Med.*, 31, S77, 2000.

222. Freitas, D. L., Maia, J. A. R., and Marques, A. T., Sexual dimorphism in physical fitness: a multivariate analysis of structural differences, in *Physical Activity and Health: Physiological, Epidemiological and Behavioral Aspects,* Gasagrande, G., and Viviani, F., Eds., Unipress, Padua, 1998, 187.

223. Fuller, N. J. and Elia, M., Potential use of bioelectrical impedance of the "whole body" and of body segments for the assessment of body composition: comparison with densitometry, *Eur. J. Clin. Nutr.,* 43, 779, 1989.

224. Gabbard, C., *Lifelong Motor Development,* Brown, Dubuque, 1992.

225. Gallahue, D. L., *Developmental Movement Experiences for Children,* J. Wiley & Sons, New York, 1982, 46.

226. Gallahue, D. L., *Understanding Motor Development: Infants, Children, Adolescents,* 2nd ed., Benchmark Press, Madison, 1989.

227. Gallahue, D., and Ozmum, J., *Understanding Motor Development,* 3rd ed., Brown and Benchmark Publishers, Madison, 1995.

228. Garn, S. M. and Lavelle, M., Two-decade follow-up of fatness in early childhood, *Am. J. Dis. Child.,* 139, 181, 1985.

229. Georgopoulos, N., Markov, K., Theodoro-Poulou, A., Paraskevopoulou, P., Varaki, L., Kazantzi, Z., Leglise, M., and Vagenakis, A. G., Growth and pubertal development in elite female rhythmic gymnasts, *J. Clin. Endocrinol. Metab.,* 84, 4525, 1999.

230. Gesell, A. and Thompson, H., *The Psychology of Early Growth, Including Norms of Infant Behaviour and a Method of Genetic Analysis,* Macmillan, New York, 1938.

231. Gilliam, T. B., Katch, V. L., Thorland, W., and Weltman, A., Prevalence of coronary heart disease risk factors in active children 7 to 12 years of age, *Med. Sci. Sports.,* 9, 21, 1977.

232. Gilliam, T. B., Freedson, P. S., Geenen, D, L., and Shahraray, B., Physical activity patterns determined by heart-rate monitoring in 6 to 7-year-old children, *Med. Sci. Sports Exercise,* 13, 65, 1981.

233. Glassow, R. B. and Kruse, P., Motor performance of girls aged 6 to 14 years, *Res. Q. Exercise Sport,* 31, 426, 1960.

234. Glover, E. G., Physical Fitness Test Items for Boys and Girls in the First, Second and Third Grades, M. S. thesis, University of North Carolina, Greensboro, 1962.

235. Godin, G. and Shephard, R. J., A simple method to assess exercise behaviour in the community, *Can. J. Appl. Sport. Sci.,* 10, 141, 1986.

236. Goldstein, H., Measuring the stability of individual growth patterns, *Ann. Hum. Biol.,* 8, 549, 1981.

237. Goran, M., I., Measurement issues related to studies of childhood obesity: assessment of body composition, body fat distribution, physical activity, and food intake, *Pediatrics,* 101, 505, 1998.

238. Goran, M. I., Kaskoun, M. C., Carpenter, W. H., Poehlman, E. T., Racressin, E., and Fontereille, A. M., Estimating body composition in young children using bioelectrical resistance, *J. Appl. Physiol.,* 75, 1776, 1993.

239. Goran, M. I., Carpenter, W. H., McGloin, A., Johnson, R., Hardin, M., and Weinsier, R. L., Energy expenditure in children of lean and obese parents, *Am. J. Physiol.,* 268, E 917, 1995.

240. Goran, M. I., Driscoll, P., Johnsson, R., Nagy, T. R., and Hunter, G. R., Cross-calibration of body composition techniques against dual-energy x-ray absorptiometry in young children, *Am. J. Clin. Nutr.*, 63, 299, 1996.

241. Goran, M. I., Hunter, G., Nagy, T. R., and Johnson, R., Physical activity related energy expenditure and fat mass in young children, *Int. J. Obesity*, 21, 171, 1997.

242. Graham, G., Motor skill acquisition: an essential goal of physical education programs, *J. Phys. Educ. Recreation Dance*, 58, 44, 1987.

243. Grasselt, W., Forchel, I., and Stemmler, R., *Zur körperlichen Entwicklung der Schuljugend in der Deutschen Demokratischen Republik*, Leipzig, 1985.

244. Graves, J. E., Pollack, M. L., Colerin, A. B., van Loan, M., and Lohman, T.G., Comparison of different bioelectrical impedance analyzers in the prediction of body composition, *Am. J. Hum. Biol.*, 1, 603, 1989.

245. Greendorfer, S. L., and Lewko, J. H., Role of the family members in sport socialization of children, *Res. Q. Exercise Sport*, 49, 146, 1978.

246. Greulich, W. W. and Pyle, S. I., *Radiographic Atlas of Skeletal Development of the Hand and Wrist*, 2nd ed., Stanford University Press, Stanford, 1959.

247. Grieve, C. and Henneberg, M., Statistical significance of body impedance measurements in estimating body composition, *Homo*, 49, 1, 1998.

248. Guo, S., Roche, A. F., and Houtkooper, L., Fat-free mass in children and young adults predicted from bioelectric impedance and anthropometric variables, *Am. J. Clin. Nutr.*, 50, 435, 1989.

249. Gutin, B., Islam, S., Manos, T., Cucuzzo, N., Smith, C., and Stachura, M. E., Relation of percentage of body fat and maximal aerobic capacity to risk factors for atherosclerosis and diabetes in black and white 7- to 11-year-old children, *J. Pediatr.*, 125, 847, 1994.

250. Gutin, B., Litauer, M., Islam, S., Manos, T., Smith, C., and Treiber, F., Body-composition measurement in 9- to 11-year-old children by dual-energy x-ray absorptiometry, skinfold-thickness measurements, and bioimpedance analysis, *Am. J. Clin. Nutr.*, 63, 287, 1996.

251. Gutin, B. S., Owens, S., Slavens, G., Riggs, S., and Treiber, F., Effect of physical training on heart-period variability in obese children, *J. Pediatr.*, 130, 938, 1997

252. Guting, G., Fogle, R. K., and Stewart, K., Relationship among submaximal heart rate, aerobic power, and running performance in children, *Res. Q Exercise Sport.*, 47, 536, 1976.

253. Hall, E. G. and Lee, A. M., Sex differences in motor performance of young children: fact or fiction, *Sex Roles*, 10, 217, 1984.

254. Halopainen, S., Lumiaho-Häkkinen, P., and Telama, R., Level and rate of development of motor fitness, motor abilities and skills by somatotype, *Scand. J. Sports Sci.*, 6, 67, 1984.

255. Halverson, L. E., Roberton, M. A., and Langendorfer, S., Development of the overarm throw: movement and ball velocity changes by seventh grade, *Res. Q. Exercise Sport*, 53, 198, 1982.

256. Halverson, L. E. and Williams, K., Developmental sequences for hopping over distance: a prelongitudinal screening, *Res. Q. Exercise Sport*, 56, 37, 1985.

257. Hansen, L., Klausen, K., Bangsbo, J., and Müller, J., Short longitudinal study of boys playing soccer: parental height, birth weight and length, anthropometry, and pubertal maturation in elite and non-elite players, *Pediatr. Exercise Sci.*, 11, 199, 1999.

258. Hardin, D. H. and Garcia, M. J., Diagnostic performance tests for elementary children (grades 1-4), *JOPERD*, 53, 48, 1982.

259. Hardman, K. and Marshall, J. J., *Worldwide Survey of the State and Status of School Physical Education, Summary of Findings.*, University of Manchester, Manchester, 1999.

260. Harlan, W. R., Cornoni-Huntley, J., and Leaverton, P. E., Blood pressure in childhood: the national health examination survey, *Hypertension*, 1, 559, 1979.

261. Harscha, D. V., The benefits of physical activity in childhood, *Am. J. Med. Sci.*, 310, S109, 1995.

262. Hartley, G. A., Comparative view of talent selection for sport in two socialist states — the USSR and the GDR — with particular reference to gymnastics, in *The Growing Child in Competitive Sport*, The National Coaching Foundation, Leeds, 1988, 50.

263. Haschke, F., Body composition of adolescent males, Part 1. Total body water in normal adolescent males, *Acta Paediatr. Scand.*, 307, S1, 1983.

264. Haschke, F., Body composition of adolescent males, Part 2. Body composition of male reference adolescents, *Acta Paediatr. Scand.*, 307, S13, 1983.

265. Haschke, F., Fomon, S. J., and Ziegler, E. E., Body composition of a nine-year old reference boy, *Pediatr. Res.*, 15, 847, 1981.

266. Haskell, W. L., Montoye, H. J., and Quenstein, D., Physical activity and exercise to achieve health-related physical fitness components, *Publ. Health Rep.*, 100, 202, 1985.

267. Haskell, W. L., Leon, A. S., Caspersen, C. J., Froelicher, V. F., Hegberg, J. M., Harlan, W., Holloszy, J. O., Regensteiner, J. G., Thompson, P. D., Washburn, R. A., and Wilson, P. W. F., Cardiovascular benefits and assessment of physical activity and physical fitness in adults, *Med. Sci. Sports Exercise*, 24, 520, 1992.

268. Haubenstricker, J. L., Branta, C. F., and Seefeldt, V. D., Standards of Performance for Throwing and Catching. Paper Presented at the Annual Conference of the North American Society for the Psychology of Sport and Physical Activity, East Lansing, MI, 1983.

269. Haubenstricker, J., and Seefeldt, V., Acquisition of motor skills during childhood, in *Physical Activity and Well-Being*, Seefeldt, V., Ed., American Alliance for Health, Physical Education, Recreation and Dance, Reston, 1986, 41.

270. Hayden, F. J. and Yuhasz, M., *The CAHPER Test Manual for Boys and Girls 7 to 17 years of Age*, Canadian Association for Health, Physical Education and Recreation, Toronto, 1966.

271. Hayes, P. A., Sowwood, P. J., Belyavin, A., Cohen, J. B., and Smith, F. W., Subcutaneous fat thickness measurement by magnetic resonance imaging, ultrasound and calipers, *Med. Sci. Sports Exercise*, 20, 303, 1988.

272. Haywood, K. M., *Life Span Motor Development*, Human Kinetics, Champaign, 1993.

273. Heath, B. H. and Carter, J. E. L., A modified somatotype method, *Am. J. Phys. Anthropology*, 24, 87, 1967.

274. Hebbelinck, M. and Borms, J., *Tests en Normschalen [Tests and Norm Scales]*, Free University, Brussels, 1969.

275. Hebbelinck, M., Duquet, W., Borms, J., and Carter, J. E. L., Stability of somatotypes: a longitudinal growth study in Belgian children followed from 6 to 17 years, *Am. J. Hum. Biol.*, 7, 575, 1995.

276. Hellebrandt, F. A., Rarick, G. L., Glassow, W., and Carns, M. L., Physiological analysis of basic motor skills I. Growth and development of jumping, *Am. J. Phys. Med.*, 40, 14, 1961.
277. Hewitt, M. J., Going, S. B., Williams, D. P., and Lohman, T. G., Hydration of fat-free body in children and adults: implications for body composition assessment, *Am. J. Physiol.*, 265, E88, 1993.
278. Heyward, V. H., Practical body composition assessment for children, adults and older adults, *Int. J. Sport. Nutr.*, 8, 285, 1998.
279. Heyward, V. H. and Stolarczyk, L. M., *Applied Body Composition Assessment*, Human Kinetics, Champaign, 1996.
280. Hill, G. M. and Miller, T. A., A comparison of peer and teacher assessment of students physical fitness performance, *Phys. Educator*, 54, 40, 1997.
281. Hillman, M., One false move … an overview of the findings and issues they raise, in *Children, Transport and the Quality of Life*, Hillman, M., Ed., Policy Studies Institute, London, 1993, 7.
282. Hills, A. P. and Parker, A. W., Obesity management via diet and exercise intervention, *Child Care Health Dev.*, 14, 409, 1988.
283. Hills, A. P., Lyell, L., and Byrne, N. M., An evaluation of the methodology for the assessment of body composition in children and adolescents, in Body Composition Assessment in Children and Adolescents, Jürimäe, T., and Hills, A. P., Eds., *Med. Sport Sci.*, Karger, Basel, 44, 1, 2001.
284. Hoffer, E. D., Meador, C. K., and Simpson, D. C., Correlation of whole-body impedance with total body water, *J. Appl. Physiol.*, 27, 531, 1969.
285. Holland, B. V., Development and Validation of an Elementary Motor Performance Test for Students Classified as Non-Handicapped, Learning Disabled or Educable Mentally Impaired, doctoral dissertation, Michigan State University, 1986.
286. Housner, L. D., Sex-role stereotyping: implications for teaching elementary physical education, *Motor Skills: Theory into Practice*, 5, 107, 1981.
287. Houtkooper, L. B., Assessment of body composition in youths and relationship to sport, *Int. J. Sport. Nutr.*, 6, 146, 1996.
288. Houtkooper, L. B., Lohman, T. G., Going, S. B., and Hall, M. C., Validity of bioelectric impedance for body composition assessment in children, *J. Appl. Physiol.*, 66, 814, 1989.
289. Houtkooper, L. B., Going, S. B., Lohman, T. G., Roche, A. F., and Van Loan, M., Bioelectrical impedance estimation of fat-free body mass in children and youth: a cross-validation study, *J. Appl. Physiol.*, 72, 366, 1992.
290. Houtkooper, L. B., Lohman, T. G., Going, S. B., and Hawell, W. H., Why bioelectrical impedance analysis should be used for estimating adiposity, *Am. J. Clin. Nutr.*, 64, 436S, 1996.
291. Illingworth, R. S. and Lister, J., The critical or sensitive periods, with special reference to certain feeding problems in infants and children, *J. Pediatr*, 65, 839, 1964.
292. Institute for Aerobics Research. *Fitnessgram User's Manual*, Dallas, 1987.
293. Institute for Aerobics Research. *Fitnessgram*, Dallas, 1988.
294. Institute for Aerobics Research, *Fitnessgram*, Dallas, 1989.
295. Isaacs, L. D., Effects of ball size, ball color, and preferred color on catching by young children, *Percept. Motor Skills*, 51, 583, 1980.

296. Ishiko, T., Merits of various standard test protocols — a comparison between ICPAFR, WHO and IBP and other groups, in *Physical Fitness Assessment, Principles, Practice and Applications,* Shephard, R. J., and Lavallee, H., Charles C. Thomas, Springfield, 1978.

297. Jackson, A. and Coleman, A., Validation of distance run tests for elementary school children, *Res. Q. Exercise Sport* 47, 36, 1976.

298. Jackson, A. S. and Pollock, M. L., Factor analysis and multivariate scaling of anthropometric variables for the assessment of body composition, *Med. Sci. Sports Exercise,* 8, 196, 1976.

299. Jackson, A. S. and Pollock, M. L., Generalized equations for predicting body density of men, *Br. J. Nutr.,* 48, 497, 1978.

300. Jackson, A. S., Pollock, M. L., and Ward, A., Generalized equations for predicting body density of women, *Med. Sci. Sports Exercise,* 12, 175, 1980.

301. Jackson, A. S., Pollock, M. L., Graves, J. E., and Mahar, M. T., Reliability and validity of bioelectrical impedance in determining body composition, *J. Appl. Physiol.,* 64, 529, 1988.

302. Janz, K. F., Validation of the CSA accelerometer for assessing children's physical activity, *Med. Sci. Sports Exercise,* 26, 369, 1994.

303. Janz, K. F., Nielson, D. H., Cassady, S. L., Cook, J. S., Wu, Y. T., and Hansen, J. R., Cross-validation of the Slaughter skinfold equation for children and adolescents, *Med. Sci. Sports Exercise,* 25, 1070, 1993.

304. Janz, K. F. and Mahoney, L. T., Three-year follow-up of changes in aerobic fitness during puberty: The Muscatine study, *Med. Sci. Sports Exercise,* 28, S11, 1996.

305. Janz, K. F., Dawson, J. D., and Mahoney, L. T., Tracking physical fitness and physical activity from childhood to adolescence: the Muscatine study, *Med. Sci. Sports Exercise,* 32, 1250, 2000.

306. Jess, M. C., Collins, D., and Burwitz, L., Children and physical activity: the centrality of basic movement skill development, in *Active Living Through Quality Physical Education,* Fischer, R., Laws, C., and Moses, J., Eds., London, 1998, 90-95.

307. Johnson, E. and La Von, C., Effects of 5-day-a-week vs. 2- and 3-day-a-week physical education class on fitness, skill, adipose tissue and growth, *Res. Q. Exercise Sport,* 4, 93, 1969.

308. Jones, M. A., Hitchen, P. J., and Stratton, G., The importance of considering biological maturity when assessing physical fitness measures in girls and boys aged 10 to 16 years, *Ann. Hum. Biol.,* 27, 57, 2000.

309. Jürimäe, J., Jürimäe, T., Sööt, T., and Leppik, A., Assessment of body composition in 9- to 11-year-old children by skinfold thickness measurements and bioelectrical impedance analysis: comparison of different regression equations, *Med. Dello Sport,* 51, 341, 1998.

310. Jürimäe, J., Leppik, A., and Jürimäe, T., Whole-body resistance measured between different limbs and resistance indices in pre-adolescent children, in Body Composition Assessment in Children and Adolescents, Jürimäe, T., and Hills, A., Eds., *Med. Sport Sci.,* Karger, Basel, 44, 53, 2001.

311. Jürimäe, J. and Jürimäe, T., Unpublished data.

312. Jürimäe, T. and Jürisson, A., The relationships between physical fitness and physical activity in children, *Children and Exercise XIX,* Welsman J., Armstrong N. and Kirby B., Eds., Vol. II, Singer Press, Washington, 1997, 101.

313. Jürimäe, T. and Volbekiene, V., Eurofit test results in Estonian and Lithuanian 11- to 17-year-old children: a comparative study, *Eur. J. Phys. Educ.*, 3, 178, 1998.

314. Jürimäe, T., Jürimäe, J., and Leppik, A., Relationships between body bioelectric resistance and somatotype in pre-adolescence children, *Acta Kinesiol. Univ. Tartuensis*, 4, 103, 1999.

315. Jürimäe, T., Jürimäe, J., and Leppik, A., Relationships between bioelectric resistance and somatotype in 9- to 11-year-old children, *Ann. N.Y. Acad. Sci.*, 904, 187, 2000.

316. Jürimäe, T., Leppik, A., and Jürimäe, J., Influence of anthropometric variables to the whole-body resistance in pre-adolescent children, in Body Composition Assessment in Children and Adolescents, Jürimäe, T. and Hills, A., Eds., *Med. Sport Sci.*, Vol. 44, Karger, Basel, 44, 61, 2001.

317. Jürimäe, T. and Hills A. P., Eds., Body Composition Assessment in Children and Adolescents, *Med. Sport Sci.*, Vol. 44., Karger, Basel, 2001, (in press).

318. Jürimäe, T., Sudi, K., Jürimäe, J., Payerl, D., Tafeit, E., Möller, R., unpublished data.

319. Jürimäe, T., Veldre, G., and Jürimäe, J., unpublished data.

320. Karvonen, M. J., Physical fitness of Finnish schoolchildren, in *International Research in Sport and Physical Education*, Jokl, E., Simon, E., Eds., Charles C. Thomas, Springfield, 1964, 479.

321. Katzmarzyk, P. T., Malina, R. M., and Beunen, G. P., The contribution of biological maturation to the strength and motor fitness of children, *Ann. Hum. Biol.*, 24, 493, 1997.

322. Katzmarzyk, P. T., Malina, R. M., Song, T. M. K., and Bouchard, C., Physical activity and health-related fitness in youth: a multivariate analysis, *Med. Sci. Sports Exercise*, 30, 709, 1998.

323. Katzmarzyk, P. T., Malina, R. M., and Bouchard, C., Physical activity, physical fitness, and coronary heart disease risk factors in youth: the Quebec family study, *Prev. Med.*, 29, 555, 1999.

324. Kelly, L. E., Dagger, J., and Walkley, J., The effects of an assessment-based physical education program on motor skill development in preschool children, *Educ. Treat. Child.*, 12, 152, 1989.

325. Kemper, H. C. G., Growth, Health and Fitness of Teenagers: Longitudinal Research in International Perspective, *Med. Sport Sci.*, Karger, Basel, 20,1985.

326. Kemper, H. C. G., *The Amsterdam Growth Study. A Longitudinal Analysis of Health, Fitness, and Lifestyle*, HK Sport Science Monograph Series, Vol. 6, Human Kinetics, Champaign, 1995.

327. Kemper, H. C. G., Snel, J., Verschuur, R., and Storm-Van Essen, L., Tracking of health and risk indicators of cardiovascular diseases from teenager to adult: Amsterdam growth and health study, *Prev. Med.* 19, 642, 1990.

328. Kemper, H. C. G., Spekreijse, M., Slooten, J., Post, G. B., Welten, D. C., and Coudet, J., Physical activity in prepubescent children: relationship with residential altitude and socioeconomic status, *Pediatr. Exercise Sci.*, 8, 57, 1996.

329. Kemper, H. C. G. and van Mechelen, W., Physical fitness testing of children: a European perspective, *Pediatr. Exercise Sci.*, 8, 201, 1996.

330. Keogh, J. F., Change in Motor Performance During Early School Years, Los Angeles: Department of Physical Education, University of California, Technical Report, 1969, 2.

331. Kikuchi, S., Rona, R. J., and Chinn, S., Physical fitness of 9 year olds in England: related factors, *J. Epidemiol. Comm. Health*, 49, 180, 1995.

332. Kilanowski, C. K., Consalvi, A. R., and Epstein, L. H., Validation of an electronic pedometer for measurement of physical activity in children, *Pediatr. Exercise Sci.*, 11, 63, 1999.

333. Kim, M., and Matsuura, Y., Annual changing trend of physical fitness of Japanese in the recent 10 years (1985 to 1994), in *Physical Activity and Health: Physiological, Epidemiological and Behavioral Aspects*, Gasagrande, G., and Viviani, F., Eds., Unipress, Padua, 1998, 129.

334. Klesges, L. M. and Klesges, R. C., The assessment of children's physical activity: a comparison of methods, *Med. Sci. Sports Exercise*, 19, 511, 1987.

335. Klesges, R. C., Klesges, L. M., Swenson, A. M., and Pheley, A. M., A validation of two motion sensors in the prediction of child and adult physical activity levels, *Am. J. Epidemiol.*, 122, 400, 1985.

336. Klish, W. J., Methods for measuring body composition in infants and children: the gold standard, in *Body Composition Measurements in Infants and Children. Report of the 98th Ross Conference on Pediatric Research*, Ross Laboratories, Columbus, 1989, 4.

337. Klish, W. J., Childhood obesity: pathophysiology and treatment, *Acta Paediatr. Jpn.*, 37, 1, 1995.

338. *"Klug & Fit"*, *Sportmotorische Tests. Schulärztliche Untersuchung des Bewegungsapparates Muskelfunktionsprüfung*, Bad Vöslav, Wien, 1994.

339. Knuttgen, H. G., Aerobic capacity of adolescents, *J. Appl. Physiol.*, 22, 655, 1967.

340. Kobayashi, K., Kitamura, K., Mikra, M., Sodeyama, H., Murase, Y., Miyashita, M., and Matsui, H., Aerobic power as related to body growth and training in Japanese boys: a longitudinal study, *J. Appl. Physiol.*, 44, 666, 1978.

341. Kodlin, D. and Thompson, G. D., An appraisal of the longitudinal approach to studies in growth and development, *Monogr. Soc. Res. Child Dev.*, 23, 7, 1958.

342. Kohl, H. W. and Hobbs, K. E., Development of physical activity behaviors among children and adolescents, *Pediatrics*, 101, 549, 1989.

343. Koo, M. M. and Rohan, T. E., Comparison of four habitual physical activity questionnaires in girls aged 7 to 15 years, *Med. Sci. Sports Exercise*, 31, 421, 1999.

344. Koslow, R. E., Can physical fitness be a primary objective in a balanced physical education program? *J. Phys. Educ. Recreation Dance*, 59, 75, 1988.

345. Kovar, R., *Human Variation in Motor Abilities and its Genetic Basis*, Charles University, Prague, 1981, 178.

346. Krahenbuhl, G. S., Individual differences and the assessment of youth fitness, in *Encyclopedia of Physical Education, Fitness, and Sports*, Stull, G. A. and Cureton, T. K., Eds., Brighton, Salt Lake City, 1980, 470.

347. Krahenbuhl, G., Pangrazi, R., Burkett, L., Schneider, L., and Peterson, G., Field estimation of VO_{2max} in children eight years of age, *Med. Sci. Sports*, 9, 37, 1977.

348. Krahenbuhl, G. S., Skinner, J. S., and Kohrt, W. M., Developmental aspects of maximal aerobic power in children, *Exercise Sports Sci. Rev.*, 12, 503, 1985.

349. Kraus, H. and Hirschland, R. P., Minimum muscular fitness tests in school children, *Res. Q. Exercise Sport*, 25, 178, 1954.

350. Kraus, H. and Raab, V., *Hypokinetic Disease*, Charles C. Thomas, Springfield, 1961.
351. Krombholz, H., Physical performance in relation to age, sex, social class and sports activities in kindergarten and elementary school, *Percept. Motor Skills*, 84, 1168, 1997.
352. Kuczmarski, R. J., Flegal, K. M., Campbell, S. M., and Johnson, C. L., Increasing prevalence of overweight among U.S. adults, *JAMA*, 272, 205, 1994.
353. Kuhlman, J. S. and Beitel, P. A., *Pattern of Relationships of Coincidence Anticipation with Age, Gender, and Depth of Sport Experience*, The University of Tennessee, Knoxville.
354. Kuntzleman, C. T. and Reiff, G. G., The decline in American children's fitness levels, *Res. Q. Exercise Sport*, 63, 107, 1992.
355. Kuschner, R. F., Bioelectrical impedance analysis: a review of principles and applications., *J. Am. Coll. Nutr.*, 11, 199, 1992.
356. Kwee, A. and Wilmore, J. H., Cardiorespiratory fitness and risk factors for coronary artery disease in 8- to 15-year old boys, *Pediatr. Exercise Sci.*, 2, 372, 1990.
357. Laakso, L. and Telama, R., Ecological differences in organized and unorganized sport participation among Finnish youth, in *Research Institute of Physical Culture and Health Yearbook*, Haajanen, T., Ed., Jyväskylä, 1981, 84.
358. Langendorfer, S., Longitudinal Evidence for Developmental Changes in the Preparatory Phase of the Overarm Throw for Force, paper presented at the Research Section of the American Alliance for Health, Physical Education, Recreation and Dance, Detroit, 1980.
359. LaPorte, R. E., Montoye, H. J., and Caspersen, C. J., Assessment of physical activity in epidemiological research: problems and prospects, *Publ. Health Rep.*, 100, 131, 1985.
360. Laws, C., The Fellows lecture 1995, opportunities for partnerships between PE and Sport, *Br. J. Phys. Educ.*, 27, 8, 1996.
361. Lefevre, J., Reference values and norms for Belgian primary schoolchildren, in *The Eurofit of Physical Fitness*, Ismir, 1990, 125.
362. Lefevre, J., Beunen, G., Borms, J., Renson, R., Vrijens, J., Claessens, A. C., and Van der Aerschot, H., *Eurofit: Leidraad bij de Testafneming en Referentiewaarden voor 6-tot en met 12-Jarige Jogens en Meisjes [Eurofit: Guideline for Testing and Reference Values of 6- to 12-year-old boys and girls]* (Monograph), Lichamelijke Opvoeding, 22, 1993, 105.
363. Lefevre, J., Beunen, G., Borms, J., and Vrijens, J., Sex differences in physical fitness in Flemish youth, in Physical Fitness and Nutrition during Growth, Parizkova, J. and Hills, A. P., Eds., *Med. Sport Sci.*, Karger, Basel, 43, 1998, 54.
364. Lewko, J. H. and Greendorfer, S. L., Family influences and sex differences in children´s socialization into sport: a review, in *Children in Sport*, 3rd ed., Smoll, F. L., Magill, R. A., and Ash, M. J., Eds., Human Kinetics, Champaign, 1982, 265.
365. Leyten, C., *De MOPER Fitheids Test, Onderzoeksveslag 9 t/m 11 Jarigen [The MOPER Fitness Test Report 9 - 11 Year Old]*, Haarlem, De Vrieseborch, 1981.
366. Liemohn, W., Choosing the safe exercise, *ACSM Certified News*, 2, 1, 1991.
367. Linder, C. W. and Durant, R. H., Exercise, serum lipids, and cardiovascular disease risk factors in children, *Pediatr. Clin. North Am.*, 29, 1314, 1982.

368. Lindgren, G., Socio-economic background, growth, educational outcome and health, in *Essays on Auxology,* Hauspie, R., Lindgren, G., and Falkner, F., Eds., Castlehead Publications, Ware, 1995, 408.

369. Lindquist, C. H., Reynolds, K. D., and Goran, M. I., Sociocultural determinants of physical activity among children, *Prev. Med.,* 29, 305, 1999.

370. Lintunet, T., Development of self-perceptions during the school years, in *Psychology for Physical Educators,* Auweele, Y. V., Bakker, F., Biddle, S., Durand, M., and Seiler, R., Eds., Human Kinetics, Champaign, 1999, 115.

371. Little, N. G., Day, J. A. P., and Steinke, L., Relationship of physical performance to maturation in perimenarcheal girls, *Am. J. Hum. Biol.,* 9, 163, 1997.

372. Livingstone, M. B. E., Coward, W. A., Prentice, A. M., Davies, P. S. W., Strain, J. J., McKenna, P. G., Mahoney, C. A., Whete, J. A., Stewart, C. M., and Kerr, M. J., Daily energy expenditure in free-living children: comparison of heart-rate monitoring with the doubly labeled water ($^2H_2^{18}O$) method, *Am. J. Clin. Nutr.,* 56, 343, 1992.

373. Lohman, T. G., Skinfolds and body density and their relation to body fatness: a review, *Hum. Biol.,* 53, 181, 1981.

374. Lohman, T. G., Research progress in validation of laboratory methods of assessing body composition, *Med. Sci. Sports Exercise,* 16, 596, 1984.

375. Lohman, T. G., Applicability of body composition techniques and constants for children and youth, in *Exercise and Sport Sciences Reviews,* Pandolf, K. B., Ed., MacMillan, New York, 1986, 325.

376. Lohman, T. G., Assessment of body composition in children, *Pediatr. Exercise Sci.,* 1, 19, 1989.

377. Lohman, T. G., *Advances in Body Composition Assessment, Current Issues in Exercise Science Series, Monograph No 3,* Human Kinetics, Champaign, 1992.

378. Lohman, T. G., Boileau, R. A., and Slaughter, M. H., Body composition in children and youth, in *Advances in Pediatric Sport Sciences,* Boileau, R. A., Ed., Human Kinetics, Champaign, 1984, 29.

379. Lohman, T. G., Pollock, M. L., Slaughter, M. H., Branden, L. J., and Boileau, R. A., Methodological factors and the prediction of body fat in female athletes, *Med. Sci. Sports Exercise,* 15, 92, 1984.

380. Lohman, T. G., Slaughter, M. H., Boileau, R. A., Bunt, J., and Lussier, L., Bone mineral measurements and their relation to body density in children, youth and adults, *Hum. Biol.,* 56, 667, 1984.

381. Lohman, T. G., Roche, A. F., and Martorell, R., *Anthropometric Standardization Reference Manual,* Human Kinetics, Champaign, 1988.

382. Lohman, T. G., Going, S. B., Slaughter, M. H., and Boileau, R. A., Concept of chemical immaturity in body composition estimates: implications for estimating the prevalence of obesity in childhood and youth, *Am. J. Hum. Biol.,* 1, 201, 1989.

383. Loko, J., Aule, R., Sikkut, T., Ereline, J., and Viru, A., Motor performance status in 10- to 17-year-old Estonian girls, *Scand. J. Med. Sports,* 10, 109, 2000.

384. Loovis, E. M. and Butterfield, S. A., Influence of age, sex, balance, and sport participation on development of catching by children grades K-8, *Percept. Motor Skills,* 77, 1267, 1993.

385. Lopes, V. P. and Maia, J. A. R., Physical education and the development of physical fitness in children, *Proc. of the XIXth Int. Symp. of the Europ. Group of Pediatr. Work Physiol.*, Vol. II, Welsman, J., Armstrong, M., and Kirbi, B., (Eds.), Singer Press, Exeter, 1997, 61.
386. Lukaski, H. C., Assessment for body composition using tetrapolar bioelectrical impedance analysis, in *New Techniques in Nutritional Research*, Academic Press, New York, 1991, 303.
387. Lukaski, H. C., Johnson, P. E., Bolonchuk, W. W., and Lykken, G. I., Assessment of fat-free mass using bioelectric impedance measurements of the human body, *Am. J. Clin. Nutr.*, 41, 810, 1985.
388. Lukaski, H. C., Bolonchuk, W. W., Hall, C. B., and Siders, W., Validation of tetrapolar bioelectrical impedance method to assess human body composition, *J. Appl. Physiol.*, 60, 1327, 1986.
389. Luke, M. D. and Sinclair, G. D., Gender differences in adolescents' attitudes toward school physical education, *J. Teachers Phys. Educ.*, 11, 31, 1991.
390. Maccoby, E. E. and Jacklin, C. N., *The Psychology of Sex Differences*, Stanford University Press, Stanford, 1974.
391. Maes, H., Beunen, G., Vlietinck, R., Lefevre, J., Van Den Bossche, C., Claessens, A., Derom, R., Lysens, R., Renson, R., Simons, J., and Van Den Eynde, B., Heritability of Health- and Performance-Related Fitness, Data from the *Leuven Longitudinal Twin Study. Kinanthropometry IV*, Duquet, W., and Day, J. A. P., Eds., E. & F.N. Spon, London, 1993.
392. Maes, H. H. M., Beunen, G. P., Vlietinck, R. F., Neale, M. C., Thomis, M., Van Den Eynde, B., Lysens, R., Simons, J., Derom, C., and Derom, R., Inheritance of physical fitness in 10-year-old twins and their parents, *Med. Sci. Sports Exercise*, 28, 1479, 1996.
393. Maffies, C., Zaffanello, M., and Schutz, Y., Relationship between physical inactivity and adiposity in prepubertal boys, *J. Pediatr.*, 131, 288, 1997.
394. Mafulli, M., Children in sport: toward the year 2000, *Sports Exercise Injury*, 63, 96, 1996.
395. Mahon, A., Ignico, A., and March, M. L., The effects of daily physical education on the health related physical fitness in first-grade children, *Res. Q. Exercise Sport*, 64, 43, 1993.
396. Maksud, M. G. and Coutts, K. D., Application of the Cooper twelve-minute run-walk test of young males, *Res. Q. Exercise Sport*, 42, 54, 1971.
397. Malina, R. M., Anthropometric correlates of strength and motor performance, *Exercise Sport Sci. Rev.*, 3, 249, 1975.
398. Malina, R. M., Growth, maturation and human performance, in *Perspectives on the Academic Discipline of Physical Education*, Brooks, G. A., Ed., Human Kinetics, Champaign, 1981, 190.
399. Malina, R. M., Readiness for competitive sport, in *Sport for Children and Youths*, Weiss, M. R. and Gould, D., Eds., Human Kinetics, Champaign, 1986, 45.
400. Malina, R. M., Readiness for competitive sport, in *The Growing Child in Competitive Sport*, The National Coaching Foundation, Leeds, 1988, 67.
401. Malina, R. M., Growth, exercise, fitness, and later outcomes, in *Exercise, Fitness, and Health: a Consensus of Current Knowledge*, Bouchard, C., Shephard, R. J., Stephens, J. R., and McPherson, B. D., Eds., Human Kinetics, Champaign, 1990, 637.

402. Malina, R. M., Physical activity: relationship to growth, maturation, and physical fitness, in *Physical Activity, Fitness, and Health,* Bouchard, C., Shephard, R. J., and Stephens, T., Eds., Human Kinetics, Champaign, 1994, 918.

403. Malina, R. M., Physical growth and biological maturation of young athletes, *Exercise Sport Sci. Rev.,* 22, 389, 1994.

404. Malina, R. M., Physical activity and fitness of children and youth: questions and implications, *Med. Exercise Nutr. Health.,* 4, 125, 1995.

405. Malina, R. M., Tracking of physical activity and physical fitness across the lifespan, *Res. Q. Exercise Sport,* 67, S48, 1996.

406. Malina, R. M. and Rarick, G. L., Growth, physique, and motor performance, in *Physical Activity: Human Growth and Development,* Rarick, G. L., Ed., Academic Press, New York, 1973, 125.

407. Malina, R. M. and Bouchard, C., *Growth, Maturation and Physical Activity,* Human Kinetics, Champaign, 1991.

408. Malina, R. M., Beunen, G. P., Claessens, A. L., Lefevre, J., Van Den Eynde, B. V., Renson, R., Vanreusel, B., and Simons, J., Fatness and physical fitness of girls 7 to 17 years, *Obesity Res.,* 3, 221, 1995.

409. Malina, R. M., and Beunen, G., Monitoring of growth and maturation, in *The Child and Adolescent Athlete,* Bar-Or, O., Ed., Blackwell Science, Oxford, 1996, 647.

410. Malina, R. M., Woynarowska, B., Bielicki, T., Beuneg, G., Eweld, D., Geithner, C. A., Yi-Ching Huang, and Rogers, D. M., Prospective and retrospective longitudinal studies of the growth, maturation, and fitness of Polish youth active in sport, *Int. J. Sports Med.* 18 (Suppl. 3), S179, 1997.

411. Manitoba Education, *Manitoba Physical Fitness Test Manual and Fitness Objectives,* Winnipeg, 1977.

412. Marschall, S. J., Sarkin, J. A., Sallis, J. F., and McKenzie, T. L., Tracking of health-related fitness components in youth ages 9 to 12, *Med. Sci. Sports Exercise,* 30, 910, 1998.

413. Marshall, W. A., Interrelationships of skeletal maturation, sexual development and somatic growth in man, *Ann. Hum. Biol.,* 1, 29, 1974.

414. Matejkova, J., Koprivova, Z., and Placheta, Z., Changes in acid-base balance after maximal exercise, in *Youth and Physical Activity,* Z. Placheta., Ed., Purkyne University, Brno, 1980, 191.

415. Mathews, D. K., Shaw, W., and Woods, J. B., Hip flexibility of elementary school boys as related to body segments, *Res. Q. Exercise Sport,* 30, 297, 1959.

416. Matthews, C. E. and Freedson, P. S., Field trial of a three-dimensional activity monitor: comparison with self-report, *Med. Sci. Sports Exercise,* 27, 1071, 1995.

417. Mazess, R. B. and Cameron, J. R., Skeletal growth in school children: maturation and bone mass, *Am. J. Phys. Anthropol.,* 35, 399, 1971.

418. Mazess, R. B., Barden, H. S., Bisek, J. P., and Hansen, J., Dual-energy x-ray absorptiometry for total body and regional bone mineral soft tissue components, *Am. J. Clin. Nutr.,* 51, 1106, 1990.

419. McKay, H. A., Bailey, D. A., Wilkinson, A. A., and Houston, C. S., Familial comparison of bone mineral density at the proximal femur and lumbar spine, *Bone Mineral,* 24, 95, 1994.

420. McKenzie, T. L., Sallis, J. F., Faucetter, N. F., Roby, J. J., and Kolody, B., Effects of a curriculum and in-service program on the quantity and quality of elementary physical education classes, *Res. Q. Exercise Sport*, 64, 25, 1993.

421. McMurray, R. G., Bradley, C. B., Harrell, J. S., Bernthal, P. R., Frauman, A. C., and Bangdiwala, S. I., Parental influences on childhood fitness and activity patterns, *Res. Q. Exercise Sport*, 64, 249, 1993.

422. McNaughton, L., Morgan, R., Smith, P., and Hannan, G., An investigation into the fitness levels of Tasmanian primary schoolchildren, *ACHPER Healthy Lifestyles J.*, 43, 4, 1996.

423. Mechelen, W., The construction of Eurofit reference scales in the Netherlands schoolchildren, in *The Eurofit Tests of Physical Fitness*, Ismir, 1990, 193.

424. Mechelen, W., van Lier, W. H., Hlobil, M., Crolla, I., and Kemper, H. C. G., *Handbook with Reference Scales for 12- to 16-Year-Old Boys and Girls in the Netherlands*, De Vrieseborch, 1991.

425. Mekota, K. and Kovar, R., *Unifittest (6-60)*, Prague, 1996.

426. Melanson, E. L., Freedson, P. S., Hendelman, D., and Debold, E., Reliability and validity of a portable metabolic measurement system, *Can. J. Appl. Physiol.*, 21, 109, 1996.

427. Melanson, E. L. and Freedson, P. S., Physical activity assessment: a review of methods, *Crit. Rev. Food Sci. Nutr.*, 36, 385, 1996.

428. Mercier, B., Mercier, J., Granier, P., Le Gallais, D., and Prefaut, C., Maximal anaerobic power: relationship to anthropometric characteristics during growth, *Int. J. Sports Med.*, 13, 21, 1992.

429. Mero, A., Jaakkola, L., and Komi, P. V., Serum hormones and physical performance capacity in young boy athletes during 1-year training period, *Eur. J. Appl. Physiol.*, 60, 32, 1990.

430. Mersch, F. and Stoboy, H., Strength training and muscle hypertrophy in children, in *Children and Exercise XIII*, Oseid, S., and Carlsen, K. H., Eds., Human Kinetics, Champaign, 1989, 165.

431. Messick, J. A., Prelongitudinal screening of hypothesized developmental sequences for the overhand tennis serve in experienced tennis players 9 to 19 years of age, *Res. Q. Exercise Sport*, 62, 249, 1991.

432. Micozzi, M. S., Albanes, D., Jones, D. Y., and Chumlea, W. C., Correlations of body mass indices with height, stature, and body composition in men and women in NHANES I and II, *Am. J. Clin. Nutr.*, 44, 725, 1986.

433. Milne, C., The youth fitness controversy in America, *Hum. Biol. Budapest*, 25, 459, 1994.

434. Milne, C., Seefeldt, V., and Reuschlein, P., Relationship between grade, sex, race, and motor performance in young children, *Res. Q. Exercise Sport*, 47, 726, 1976.

435. Mingda, C. and Asami, T., *Chino-Japanese Cooperative Study on Physical Fitness of Children and Youth, I*, 1986.

436. Mirwald, R. L., Bailey, D. A., Cameron, N., and Rasmusen, R. L., Longitudinal comparison of aerobic power in active and inactive boys aged 7.0 to 17.0 years, *Ann. Hum. Biol.*, 8, 405, 1981.

437. Möller, R., Tafeit, E., Smolle, K. H., and Kulling, P., Lipometer: determining the thickness of a subcutaneous fatty layer, *Biosens. Bioelectron.*, 9, 13, 1994.

438. Montoye, H. J., Kemper, H. C. G., Saris, W. H. M., and Washburn, R. A., *Measuring Physical Activity and Energy Expenditure,* Human Kinetics, Champaign, 1996.
439. Moravec, R., Kampmiller, T., and Sedlacek, J., *Eurofit — Physique and Motor Fitness of the Slovac School Youth,* Bratislava, 1996.
440. Morris, A. M., Williams, J. M., Atwater, A. E., and Wilmore, J. H., Age and sex differences in motor performance of 3 through 6 year old children, *Res. Q. Exercise Sport,* 53, 214, 1982.
441. Morris, J. N., Pollard, R., Everitt, M. G., Chaer, S. P. W., and Semmance, A. M., Vigorous exercise in leisure time: protection against coronary heart disease, *Lancet,* 2, 1207, 1980.
442. Morrow, J. R. and Freedson, P. S., Relationship between habitual physical activity and aerobic fitness in adolescents, *Pediatr. Exercise Sci.,* 6, 315, 1994.
443. NASPE, Developmentally Appropriate Practice for Children, A COPEC Position Paper, AAHPERD, Reston, 1992.
444. Neilinn-Lilienberg, K., Saava, M., and Tur, I., Height, weight, body mass index, skinfolds and their correlation to serum lipids and blood pressure in the epidemiological study of schoolchildren in Tallinn, in *Papers on Anthropology VII,* University of Tartu Press, Tartu, 1997, 243.
445. Nelson, J. K., Thomas, J. R., Nelson, K. R., and Abraham, P. C., Gender differences in children's throwing performance: biology and environment, Res. Q. *Exercise Sport,* 57, 280, 1986.
446. Nelson, K. R., Thomas, J. R., and Nelson, J. K., Longitudinal change in throwing performance: gender differences, *Res. Q. Exercise Sport,* 62, 105, 1991.
447. Nicholls, J., Achievement motivation: concepts of ability, subjective experience, task choice, and performance, *Psychol. Rev.,* 91, 328, 1984.
448. Norton, K. I., Whittingham, N., Carter, J. E. L., Kerr, D., Gore, C., and Marfell-Jones, M. J., Measurement techniques in anthropometry, in *Anthropometrica,* Norton, K. I. and Olds, T. S., Eds., UNSW Press, Sydney, 1996, 25.
449. Nupponen, H., Evaluation of the physical fitness of Finnish schoolchildren, *FIEP Bull.,* 46, 51, 1976.
450. Nupponen, H., *Koululaistel fyysis - motorinen kunto (The Physical-Motor Fitness of School Children),* Liikunnan ja Kansanterveyden Edistämissaatiö, Jyväskylä, 1981 (in Finnish).
451. Nupponen, H., *Development of Motor Abilities and Physical Activity in School Children Aged 9-16 Years,* LIKES-Research Center for Sport and Health Sciences, Jyväskylä, 1997, 326.
452. Nupponen, H., Soini, H., and Telama, R., *Koululaisten Kunnon ja Liikehallinnan Mittaaminen,* LIKES-tutkimuskeskus, Jyväskylä, 1999 (in Finnish).
453. O'Hara, N. M., Baranowski, T., Simons-Morton, B. G., Wilson, B. S., and Parcel, G. S., Validity of the observation of children's physical activity, *Res. Q. Exercise Sport,* 60, 42, 1989.
454. Oja, L. and Jürimäe, T., Assessment of motor ability of 4- and 5-year-old children, *Am. J. Hum. Biol.,* 9, 659, 1997.

455. Oja, L. and Jürimäe, T., The relationships between somatic development and fundamental motor skill performance in 6-year-old children, in *Papers on Anthropology VII*, University of Tartu Press, Tartu, 1997, 269.

456. Oja, L. and Jürimäe, T., Relationships between physical activity, motor ability, and anthropometric variables in 6-year-old Estonian children, in *Physical Fitness and Nutrition During Growth*, Parizkova, J., and Hills, A. P., Eds., Karger, Basel, 1998, 68.

457. Oja, L. and Jürimäe, T., The influence of somatic development to the motor ability and fundamental motor skill performance in 6-year-old children, in *Sport Kinetics '97*, Blaser, P., Ed., Czwalina Verlag, Hamburg, 1999, 168.

458. Okasora, K., Takaya, R., Tokuda, M., Fukunaga, Y., Oguni, T., Tanaka, H., Konishi, K., and Tamai, H., Comparison of bioelectric impedance analysis and dual energy x-ray absorptiometry for assessment of body composition in children, *Pediatr. Int.*, 41, 121, 1999.

459. Orchard, T. J., Donahue, R. P., Kuller, L. H., Hodge, P. N., and Drash, A. L., Cholesterol screening in childhood: does it predict adult hypercholesterolemia? *J. Pediatr.*, 103, 687, 1983.

460. Organ, L. W., Bradham, G. B., Gore, D. T., and Lozier, S. L., Segmental bioelectrical impedance analysis: theory and application of a new technique, *J. Appl. Physiol.*, 77, 98, 1994.

461. Paffenbarger, R. S. Jr. and Hyde, R. T., Exercise in the prevention of coronary heart disease, *Prev. Med.*, 13, 3, 1984.

462. Paffenbarger, R. S. Jr., Hyde, R. T., Wing, A. L., and Steinmetz, C. H., A natural history of athletism and cardiovascular health, *JAMA*, 252, 491, 1984.

463. Paffenbarger, R. S. Jr., Hyde, R. T., Wing, A. L., and Hsieh, C., Physical activity, all-cause mortality, and longevity of college alumni, *New Engl. J. Med.*, 314, 605, 1986.

464. Panter-Brick, C., Todd, A., Baker, R., and Worthman, C., Heart rate monitoring of physical activity among village, school, and homeless Nepali boys, *Am. J. Hum. Biol.*, 8, 661, 1996.

465. Parizkova, J., Total body fat and skinfold thickness in children, *Metabolism*, 10, 794, 1961.

466. Parizkova, J., *Nutrition, Physical Activity, and Health in Early Life*, CRC Press, Boca Raton, 1996.

467. Parizkova, J. and Carter, J. E. L., Influence of physical activity on stability of somatotypes in boys, *Am. J. Phys. Anthropol.*, 44, 327, 1976.

468. Parizkova, J. and Hills, A. P., *Childhood Obesity, Prevention and Treatment*, CRC Press, Boca Raton, 2000.

469. Pate, R. R., The evolving definition of physical fitness, *Quest*, 40, 174, 1988.

470. Pate, R. R., The case for large-scale physical fitness testing in American youth, *Pediatr. Exercise Sci.*, 1, 290, 1989.

471. Pate, R. R., Health-related measures of children's physical fitness, *J. Sch. Health*, 61, 231, 1991.

472. Pate, R. R., Physical activity assessment in children and adolescents, *Crit. Rev. Food Sci. Nutr.*, 33, 321, 1993.

473. Pate, R. R., Dowda, M., and Ross, J. G., Associations between physical activity and physical fitness in American children, *Am. J. Dis. Children*, 144, 1123, 1990.

474. Pate, R. R, Burgess, M. F., Woods, J. A., Ross, J. G., and Baumgartner, T., Validity of field tests of upper body muscular strength, *Res. Q. Exercise Sport,* 64, 17, 1993.

475. Pate, R. R., Baranowski, T., Dowda, M., and Trost, S. G., Tracking of physical activity in young children, *Med. Sci. Sports Exercise,* 28, 92, 1996.

476. Pate, R. R., Trost, S. G., Dowda, M., Ott, A. E., Ward, D. S., Sanders, R., and Felton, G., Tracking of physical activity, physical inactivity, and health-related physical fitness in rural youth, *Pediatr. Exercise Sci.,* 11, 364, 1999.

477. Perusse, L., Lortie, G., Leblanc, C., Tremblay, C., and Theriault, G., Genetic and environmental sources of variation in physical fitness, *Ann. Hum. Biol.,* 14, 425, 1987.

478. Powell, K. E. and Dysinger, W., Childhood participation in organized school sports and physical education as precursors of adult physical activity, *Am. J. Prev. Med.,* 314, 605, 1987.

479. Powell, K. E., Caspersen, C. J., Koplan, J. P., and Ford, E. S., Physical activity and chronic disease, *Am. J. Clin. Nutr.,* 49, 999, 1989.

480. Prat, J. A., Cosamart, J., Balague, N., Martinez, M., Povill, J. M., Sanchez, A., Silla, D., Santigora, S., Perez, G., Riera, J., Vela, J. M., and Partero, P., *Eurofit: La Batteria Eurofit a Cataluna [Eurofit: The Physical Fitness Battery in Spain],* Secretaria General de l' Esport, Barcelona, 1993.

481. Preberg, Z., Secular growth changes in Croatia over the twentieth century, in *Secular Growth Changes in Europe,* Bodzsar, B. E. and Susanne, C., Eds., Eötvös University Press, Budapest, 1998, 75.

482. Preece, M. A., Prepubertal and pubertal endocrinology, in *Human Growth: a Comprehensive Treatise, Vol. 2, Postnatal Growth: Neurobiology,* 2nd ed., Falkner, F., and Tanner, J. M., Eds., Plenum, New York, 1986, 211.

483. President's Council of Physical Fitness and Sports, *The Presidental Physical Fitness Award Program,* Washington, 1987.

484. Printauro, S., Nagy, T. R., Duthie, C., and Goran, M. I., Cross-calibration of fat and lean measurements by dual energy x-ray absorptiometry to pig carcass analysis in the pediatric body weight range, *Am. J. Clin. Nutr.,* 63, 293, 1996.

485. Pritchard, J. E., Nowson, C. A., Stravos, B. J., Carlson, J. S., Kaymakci, B., and Wark, J. D., Evaluation of dual energy x-ray absorptiometry as a method of measurement of body fat, *Eur. J. Clin. Nutr.,* 47, 216, 1992.

486. Prokopec, M. and Bellisele, F., Adiposity in Czech children followed from 1 month of age to adulthood: analysis of individual BMI patterns, *Ann. Hum. Biol.,* 20, 517, 1993.

487. Przeweda, R., Is secular trend also present in the physical fitness of youth, in *Proceedings of the 3rd International Conference "Sport Kinetics '93,"* Osinski, W. and Starosta, W. Eds., Poznan, Warsaw, 1994, 209.

488. Public Health Service, *Healthy People 2000: National Health Promotion and Disease Prevention Objectives,* U.S. Department of Health and Human Services, Washington, 1991.

489. Puhl, J., Greaves, K., Hoyt, M., and Baranowski, T., Children's activity and calibration, *Res. Q. Exercise Sport,* 61, 26, 1990.

490. Pyke, J., *Australian Health and Fitness Survey,* ACHPER, Adelaide, 1987.

491. Quck, J. J., Menon, J., Tan, S., and Wang, B., Review of the National Physical Award (NAPFA) Norms, in *Proc. of the International Sports Science Conference 93, Singapore,* Singapore, 1993, 161.

492. Raitakari, O. T., Taimela, S., Porkka, K. V. K., Telama, R., Valimaki, I., Akerblom, H. M., and Viikari, J.S., Associations between physical activity and risk factors for coronary heart disease: the cardiovascular risk in young finns study, *Med. Sci. Sports Exercise*, 29, 1055, 1997.

493. Rarick, G. L., The emergence of the study of human motor development, in *Perspectives on the Academic Discipline of Physical Education*, Brooks, G. A., Ed., Human Kinetics, Champaign, 1969.

494. Rarick, G. L. and Smoll, F. L., Stability of growth in strength and motor performance from childhood to adolescence, *Hum. Biol.*, 39, 295, 1967.

495. Raudsepp, L., *Physical Activity, Somatic Characteristics, Fitness and Motor Skill Development in Prepubertal Children*, Ph.D. dissertation, University of Tartu, Tartu, 1996.

496. Raudsepp, L. and Jürimäe, T., Physical activity, fitness and adiposity of prepubertal girls, *Pediatr. Exercise Sci.*, 8, 259, 1996.

497. Raudsepp, L. and Jürimäe, T., Physical activity, fitness and somatic characteristics of prepubertal girls, *Biol. Sport*, 13, 55, 1996.

498. Raudsepp, L. and Jürimäe, T., Relationships between somatic variables, physical activity, fitness and fundamental skills in prepubertal boys, *Biol. Sport*, 13, 279, 1996.

499. Raudsepp, L. and Jürimäe, T., Relationships of physical activity and somatic characteristics with physical fitness and motor skill in prepubertal girls, *Am. J. Hum. Biol.*, 9, 513, 1997.

500. Reiff, G. G., Dixon, W. R., Jacoby, D., Ye, G. H., Spain, C. G., and Hunisicker, P. A., *The President's Council on Physical Fitness and Sports 1985, National School Population Fitness Survey*, University of Michigan, Ann Arbor, 1986.

501. Reilly, J. J., Assessment of body composition in infants and children, *Nutrition*, 14, 821, 1998.

502. Reilly, J. J., Wilson, J., McColl, J. H., Carmichael, M., and Durnin, J. V., Ability of bioelectric impedance to predict fat-free mass in prepubertal children, *Pediatr. Res.*, 39, 176, 1996.

503. Reilly, T. and Stratton, G., Children and adolescents in sport: physiological considerations, *Sports Exercise Injury*, 1, 207, 1995.

504. Riddoch, C., Savage, J. M., Murphy, N., Cran, G. W., and Boreham, C., Long-term health implications of fitness and physical activity patterns, *Arch. Dis. Childhood*, 66, 1426, 1991.

505. Riddoch, C. J. and Boreham, A. C., The health-related physical activity of children, *Sports Med.*, 19, 86, 1995.

506. Roberton, M. A., Stability of stage categorizations across trials: implications for the "stage theory" of overarm throw development, *J. Hum. Movement Stud.*, 4, 167, 1978.

507. Roberton, M. A., Longitudinal evidence of developmental stages in the forceful overarm throw, *J. Hum. Movement Stud.*, 4, 167, 1978.

508. Roberton, M. A., Stability of stage categorizations across trials: implications for the stage theory of overarm throw development, *J. Hum. Movement Stud.*, 3, 49, 1977.

509. Roberton, M. A., Changing motor patterns during childhood, in *The Development of Movement Control and Co-ordination*, Kelso, J. A. S. and Clark, J. E., Eds., Burgess, Minneapolis, 1982, 48.

510. Roberton, M. A., Motor development: recognizing our roots, charting our future, *Quest*, 41, 213, 1989.
511. Roberton, M. A. and Halverson, L. E., The developing child — his changing movement, in *Physical Education for Children: a Focus on the Teaching Process*, Logston, B. J., Ed., Lea & Febiger, Philadelphia, 1977, 24.
512. Roberton, M. A. and Halverson, L. E., The development of locomotor coordination: longitudinal change and invariance, *J. Motor Behav.*, 20, 197, 1988.
513. Robertson, J., Children, *Aussie Sports and Organized Junior Sport: Final Report*, University of South Australia, Magill, 1992.
514. Roche, A. F., Some aspects of the criterion methods for the measurement of body composition, *Hum. Biol.*, 59, 209, 1987.
515. Roche, A. F., *Growth, Maturation and Body Composition. The Fels Longitudinal Study 1929-1991*, Cambridge Studies in Biological Anthropology 9, Cambridge University Press, Cambridge, 1992.
516. Roche, A. F., Siervogel, R. M., Chumlea, W. C., Reed, R. B., Valadian, I., Eichorn, D., and McCammon, R. W., *Serial Changes in Subcutaneous Fat Thickness of Children and Adults*, Karger, Basel, 1982.
517. Roche, A. F., Chumlea, W. C., and Thissen, D., *Assessing of Skeletal Maturity of the Hand-Wrist: Fels Method*, Charles C. Thomas, Springfield, 1988.
518. Roche, A. F. and Guo, S., Tracking: its analysis and significance, *Hum. Biol. Budapest*, 25, 465, 1994.
519. Roche, A. F., Guo, S., and Towne, B., Opportunities and difficulties in long-term studies of growth., *Int. J. Sports Med.*, 18, S151, 1997.
520. Roemmich, J. N. and Rogel, A. D., Physiology of growth and development. Its relationship to performance in the young athlete, *Clin. Sports Med.*, 14, 483, 1995.
521. Rolland-Cachera, M. F., Prediction of adult body composition from infant and child measurements, in *Body Composition Techniques in Health and Disease*, Davies, P. S.W. and Cole, T. J., Eds., Cambridge University Press, Cambridge 1995, 100.
522. Rolland-Cachera, M. F., Cole, T., Semper, M., Tichet, J., Rossignol, C., and Charraud, A., Variations in body mass index in the French population from 0 to 87 years, in *Obesity in Europe '91*, Ailhaud, G. A, Ed., John Wiley, London, 1991, 113.
523. Rona, R. J., Secular trend of stature and body mass index (BMI) in Britain in the 20th century, in *Secular Growth Changes in Europe*, Bodzsar, B. E. and Susanne, C., Eds, Eötvös University Press, Budapest, 1998, 335.
524. Rooks, M. A., Body Mass Index in Young Dutch Adults: Its Development and the Etiology of its Development, Ph.D. dissertation, Wageningen, 1989.
525. Ross, J. G. and Gilbert, G. G., The National Children and Youth Fitness Study: a summary of findings, *JOPERD*, 56, 45, 1985.
526. Ross, J. G. and Pate, R. R., The national children and youth fitness study II: a summary of findings, *JOPERD*, 58, 51, 1987.
527. Ross, J. G., Pate, R. R., Delphy, L. A., Gold., R. S., and Suilar, M., The national children and youth fitness study II: new health-related fitness norms, *JOPERD*, 58, 66, 1987.
528. Ross, W. D., *Anthropometry in Assessing Physique Status and Monitoring Change, The Child and Adolescent Athlete*, Bar-Or, O., Ed., Blackwell Science, Oxford, 1996, 538.

529. Rowland, T. W., *Exercise and Children's Health*, Human Kinetics, Champaign, 1990.
530. Rowland, T. W., *Developmental Exercise Physiology*, Human Kinetics, Champaign, 1996.
531. Rowland, T. W., The biological basis of physical activity, *Med. Sci. Sports Exercise*, 30, 392, 1998.
532. Rowland, T. W., Vanderburgh, R., and Cunningham, L., Body size and the growth of maximal aerobic power in children: a longitudinal analysis, *Pediatr. Exercise Sci.*, 9, 262, 1997.
533. Rowlands, A. V., Eston, R. G., and Ingledew, D. K., Measurement of physical activity in children with particular reference to the use of heart rate and pedometry, *Sports Med.*, 24, 258, 1997.
534. Rowlands, A. V., Eston, R. G., and Ingledew, D. K., Relationship between activity levels, aerobic fitness, and body fat in 8- to 10-year-old children, *J. Appl. Physiol.*, 86, 1428, 1999.
535. Ruskin, H., Physical performance of schoolchildren in Israel, in *Physical Fitness Assessment, Principles, Practice and Applications*, Shephard, R. J. and Lavallee, H., Eds., Charles C. Thomas, Springfield, 1978, 273.
536. Ryke, J. E., The Australian health and fitness survey 1985: the fitness, health and physical performance of Australian school students aged 7–15 years, *Australian Council for Health, Physical Education and Recreation*, Adelaide, 1987.
537. Sääkslahti, A., Numminen, P., Niinikoski, H., Rask-Nissilä, L., Viikari, J., Tuominen, J., and Välimäki, I., Is physical activity related to body size, fundamental motor skills, and CHD risk factors in early childhood?, *Pediatr. Exercise Sci.*, 11, 327, 1999.
538. Sady, S. P., Berg, K., Beal, D., Smith, J. L., Savage, M. P., Thompson, W. H., and Nuttel, J., Aerobic fitness and serum high-density lipoprotein cholesterol in young children, *Hum. Biol.* 56, 771, 1984.
539. Safrit, M. J., *Introduction to Measurement in Physical Education and Exercise Science*, Mosby College Publishing, St. Louis, 1990.
540. Safrit, M. J., The validity and reliability of fitness tests for children: a review, *Pediatr. Exercise Sci.*, 2, 9, 1990.
541. Safrit, M. J., *Complete Guide to Youth Fitness Testing*, Human Kinetics, Champaign, 1995, 42.
542. Safrit, M. J. and Wood, T. M., The test battery reliability of the health-related physical fitness test, *Res. Q. Exercise Sport*, 58, 160, 1987.
543. Sallis, J. F., A commentary on children and fitness: a public health perspective, *Res. Q. Exercise Sport*, 58, 326, 1987.
544. Sallis, J. F., Self-report measures of children's physical activity, *J. Sch. Health*, 61, 215, 1991.
545. Sallis, J. F., Patterson, T. L., and Buono, J. J., Relation of cardiovascular fitness and physical activity to cardiovascular disease risk factors in children and adults, *Am. J. Epidemiol.*, 127, 933, 1988.
546. Sallis, J. F., Buono, M. J., Roby, J. J., Carlson, D., and Nelson, J. A., The Caltrac accelerometer as a physical activity monitor for school-age children, *Med. Sci. Sports Exercise*, 22, 698, 1990.
547. Sallis, J. F., Buono, M. J., and Freedson, P. S., Bias in estimation of caloric expenditure from physical activity in children, *Sports Med.*, 11, 203, 1991.

548. Sallis, J. F., Simons-Morton, B. G., Stone, E. J., Corbin, C. B., Epstein, L. H., Faucette, N., Ianotti, R. J., Killen, J. D., Klesges, R. C., Petray, C. K., Rowland, T. W., and Taylor, W. C., Determinants of physical activity and interventions in youth, *Med. Sci. Sports Exercise*, 24, S248, 1992.
549. Sallis, J. F., Buono, M. J., Roby, J. J., Micale, F. G., and Nelson, J. A., Seven-day recall and other physical activity self-reports in children and adolescents, *Med. Sci. Sports Exercise*, 25, 99, 1993.
550. Sallis, J. F., McKenzie, T. L., and Alcaraz, J. E., Habitual physical activity and health-related physical fitness in fourth-grade children, *Am. J. Dis. Child.*, 147, 890, 1993.
551. Sallis, J. F. and Patrick, K., Physical activity guidelines for adolescents: consensus statement, *Pediatr. Exercise Sci.*, 6, 302, 1994.
552. Sallis, J. F., Berry, C. C., Broyles, S. L., McKenzie, T. L., and Nader, P. R., Variability and tracking of physical activity over 2 years in young children, *Med. Sci. Sports Exercise*, 27, 1042, 1995.
553. Sallis, J. F., McKenzie, T. L., Alcaraz, J., Kolody, B., Fauchette N., Roby, J., and Howell, M. F., The effects of a 2-year physical education program (SPARK) on physical activity and fitness in elementary school students, *Am. J. Publ. Health*, 87, 1328, 1997.
554. Salo, P. L., Junkala, T., Kemikangas, M., Ojala, K., Suominen, H., Kröger, H., Kannas, L., and Cheng, S., Current status of leisure-time physical activity in Finnish girls and boys aged 10-13, 5th Ann. Congr. of the ECSS., Jyväskylä, 2000, 641.
555. Sanders, S. W., *Designing Preschool Movement Program*, Human Kinetics, Champaign, 1992.
556. Saris, W. H. M., The assessment and evaluation of daily physical activity in children: a review, *Acta Paediatr. Scand.*, 318, 37, 1985.
557. Saris, W. H. M., Habitual physical activity in children: methodology and findings in health and disease, *Med. Sci. Sports Exercise*, 18, 253, 1986.
558. Saris, W. H. M., Büskhorst, R. A., Gramwinkel, A. B., van Waesberghe, F., and van der Vun-Hezemans, A. M., The relationships between working performance, daily physical activity, fatness, blood lipids and nutrition in schoolchildren, in *Children and Exercise IX*, Berg, K. and Eriksen, B. O., Eds., University Park Press, Baltimore, 1980, 166.
559. Saris, W. H. M., Elvers, J. W. H., Van't Hof, M. A., and Binkhorst, R. A., Changes in physical activity of children aged 6 to 12 years, in *Children and Exercise XII*, Rutenfranz, J. and Mocellin, R., Eds., Human Kinetics, Champaign, 1983, 121.
560. Schaefer, F., Georgi, M., Zieger, A., and Scharer, K., Usefulness of bioelectric impedance and skinfold measurements in predicting fat-free mass derived from total body potassium in children, *Pediatr. Res.*, 35, 617, 1994.
561. Schell, B. and Gross, R., The reliability of bioelectrical impedance measurements in the assessment of body composition in healthy adults, *Nutr. Rep. Int.*, 36, 449, 1987.
562. Schmidt, G. J., Walkuski, J. J., and Stensel, D. J., The Singapore youth coronary risk and activity study, *Med. Sci. Sports Exercise*, 30, 105, 1998.
563. Schmucker, N., Rigauer, W., Hinrichs, W., and Trawinski, J., Motor abilities and habitual physical activity in children, in *Children and Sport*, Ilmarinen, J. and Vallimäki, I., Eds., Springler-Verlag, Berlin, 1984, 46.

564. Schnabl-Dickey, E. A., Relationships between parent's child rearing attitudes and the jumping and throwing performance of their preschool children, *Res. Q. Exercise Sport*, 48, 382, 1977.

565. Schneider, F. J., Der neue gesundheitsorientierte Fitnesstest der USA und die Ergebnisse einer Untersuchung deutscher Kinder, *Sportunterricht* 5, 173, 1986.

566. Schoeller, D. A., Ravussion, E., Schutz, Y., Acheson, D. J., Baertschi, P., and Jecquier, E., Energy expenditure by doubly labeled water: validation in humans and proposed calculation, *Am. J. Physiol.*, 250, R823, 1986.

567. Scott, J. P., *Early Experience and the Organization of Behavior*, Brooks, Cole, Belmont, 1968.

568. Scott, J. P., Critical periods in organizational processes, in *Human Growth. Vol 1, Development Biology, Prenatal Growth*, 2nd ed., Falkner, F. and Tanner, J. M., Eds., Plenum, New York, 1986, 181.

569. Seefeldt, V., Critical Learning Periods and Programs of Early Intervention, paper presented at the AAPHER Convention, Atlantic City, 1975.

570. Seefeldt, V. and Haubenstricker, J., Patterns, phases, and stages: an analytical model for the study of developmental movement, in *The Development of Movement Control and Coordination*, Kelso, J. A. S. and Clark, J. E., Eds., John Wiley & Sons, New York, 1982, 309.

571. Seefeldt, V. and Vogel, P., Physical fitness testing of children: a 30-year history of misguided efforts, *Pediatr. Exercise Sci.*, 1, 295, 1989.

572. Sergienko, L., The genetic based prognosis in sport selection, *Kinesiology*, 31, 11, 1999.

573. Shephard, R. J., *Physical Activity and Growth*, Yearbook Medical Publications, Chicago, 1982.

574. Shephard, R. J., *Fitness of a Nation, Lessons from the Canada Fitness Survey*, Karger, Basel, 1986.

575. Shephard, R. J., Lavallee, H., Rajic, K. M., Jequier, J. C., Brisson, G., and Beaucage, Radiographic age in the interpretation of physiological and anthropometrical data, in *Pediatric Work Physiology*, Borms, J. and Hebbelinck, M., Eds., Karger, Basel, 1978, 124.

576. Shephard, R. J., Volle, M., Lavallee, H., Labarre, R., Jequier, J. C., and Rajic, M., Required physical activity and academic grades — a controlled study, in *Proceedings of the 10th Pediatric Work Physiology Symposium*, Ilmarinen, J., Ed., Springer, Jontsa, 1981.

577. Shephard, R. J. and Lavallee, H., Enhanced physical education and body fat in the primary school child, *Am. J. Hum. Biol.*, 5, 697, 1993.

578. Shephard, R. J. and Lavallee, H., Impact of enhanced physical education in the prepubescent child: trois riviers revisited, *Pediatr. Exercise Sci.*, 5, 177, 1993.

579. Shephard, R. J. and Lavallee, H., Changes of physical performance as indicators of the response to enhanced physical education, *J. Sports Med. Phys. Fitness*, 34, 323, 1994.

580. Sherif, C. W. and Rattray, G. D., Psychological development and activity in middle childhood, in *Child in Sport and Physical Activity*, Albinson, J. G. and Andrews, G. M., Eds., University Park Press, Baltimore, 1997.

581. Siels, L. R. G., The relationship between measures of physical growth and gross motor performance of primary-grade school children, *Res. Q. Exercise Sport*, 22, 244, 1951.

582. Siervogel, R. M., Roche, A. F., Guo, S. M., Mukherju, D., and Chumlea, W. C., Patterns of change in weight/stature^2 from serial data for children in the Fels longitudinal growth study, *Int. J. Obesity*, 15, 479, 1991.

583. Siervogel, R. M., Maynard, L. M., Wisemandle, W. A., Roche, A. F., Guo, S. S., Chumlea, W. C., and Towne, B., Annual changes in total body fat and fat-free mass in children from 8 to 18 years in relation to changes in body mass index, *Ann. N.Y. Acad. Sci.* 904, 2000, 420.

584. Silva, P. A., Birkbeck, J., Russell, D. G., and Wilson, J., Some biological, developmental, and social correlates of gross and fine motor performance in Dunedin seven year olds, *J. Hum. Movement Studies*, 10, 35, 1984.

585. Silvennoinen, M., Relations between different kinds of physical activity and motive types among Finnish comprehensive and upper secondary school pupils, *Scand. J. Sports Sci.*, 6, 77, 1984.

586. Simons, J., Ostyn, M., Beunen, G., Ruison, G., and Van Gerven, D., Factor analytic study of the motor ability of Belgian girls age 12 to 19, in *Biochemics of Sports and Kinanthropometry*, Landry, F. and Orball, W. A., Eds., Symposia Specialists, Miami, 1978, 395.

587. Simons-Morton, B. G., O'Hara, N. M., Simons-Morton, D. G., and Parcel, G. S., Children and fitness: a public health perspective, *Res. Q. Exercise Sport*, 58, 295, 1987.

588. Simons-Morton, B. G., Parcel, G. S., O'Hara, N. M., Blair, S. N., and Pate, R. R., Health-related physical fitness in childhood: status and recommendations, *Ann. Rev. Publ. Health*, 9, 403, 1988.

589. Simons-Morton, B. G., O'Hara, N. M., Parcel, G. S., Wei Huang, A. I., Baranowski, T., and Wilson, B., Children's frequency of participation in moderate to vigorous physical activities, *Res. Q. Exercise Sport*, 61, 307, 1990.

590. Simons-Morton, B. G., Parcel, G. S., Baranowski, T., Forthofer, R., and O'Hara, N. M., Promoting physical activity and a healthful diet among children: results of a school-based intervention study, *Am. J. Publ. Health*, 81, 986, 1991.

591. Simons-Morton, B. G., Taylor, W. C., Snider, S. A., and Huang, I. W., Physical activity of fifth-grade students during physical education, *Am. J. Publ. Health*, 83, 262, 1993.

592. Singer, R. N., *Motor Learning and Human Performance — An Application to Motor Skills and Movement Behaviours*, 3rd ed., Macmillan, New York, 1980.

593. Siri, W.E., Body composition from fluid spaces and density: Analysis of methods, in *Techniques for Measuring Body Composition*, Brozek, J. and Hunschel, A., Eds., National Academy of Sciences, Washington, 1961, 223.

594. Sjöström, L., A computer tomography based multicompartment body composition technique and anthropometric prediction of lean body mass, total and subcutaneous adipose tissue, *Int. J. Obesity*, 15, 19, 1991.

595. Slaughter, M. H., Lohman, T. G., and Misner, J. E., Relationship of somatotype and body composition to physical performance in 7- to 12-year-old boys, *Res. Q. Exercise Sport*, 46, 159, 1977.

596. Slaughter, M. H., Lohman, T. G., and Boileau, R. A., Relationship of anthropometric dimensions to physical performance in children. *J. Sports Med. Phys. Fitness*, 22, 377, 1982.

597. Slaughter, M. H., Lohman, T. G., Boileau, R. A., Horswill, C. A., Stillman, R. J., Van Loan, M. D., and Bemben, D. A., Skinfold equations for estimation of body fatness in children and youth, *Hum. Biol.*, 60, 709, 1988.

598. Sloan, A. W., Physical fitness of South African compared with British and American high school children, *S. Afr. Med. J.,* 40, 688, 1966.
599. Slooten, J., Kemper, H. C. G., Post, G. B., Lujan, C., and Coudert, J., Habitual physical activity in 10- to 12-year-old Bolivian boys: the relation between altitude and socioeconomic status, *Int. J. Sports Med.,* 15, S106, 1994.
600. Smoll, F. L., Motor impairment and social development, *Am. Corrective Ther. J.,* 58, 57, 1987.
601. Sodoma, C. J., AAU Physical Fitness Program Validity for Selected Age Group Test Results, doctoral dissertation, Indiana University, 1986.
602. Spain, C., The president´s council on physical fitness and sports, *ICSSPE Bull.,* 29, 30, 2000.
603. Sports and children: consensus statement on organized sports for children, *Bull. World. Health Org.,* 76, 445, 1998.
604. Sprynarova, S. and Parizkova, J., La stabilite de differences interindividuelles des parametres morphologiques et cardiorespiratoires czez les garcons, in *Frontiers of Activity and Child Health,* Lavallee, H. and Shephard, R. J., Eds., Pelican, Quebec, 1977, 131.
605. Spurr, G. B., Reina, J. C., and Barac-Nieto, M., Marginal malnutrition in school-aged Colombian boys: metabolic rate and estimated daily energy expenditure, *Am. J. Clin. Nutr.,* 44, 113, 1986.
606. Stephens, T., Jacobs, D. R., and White, C. C., A descriptive epidemiology of leisure-time physical activity, *Publ. Health Rep.,* 100, 147, 1985.
607. Stephens, T., and Casperson, C. J., The demography of physical activity, in *Physical Activity, Fitness and Health: International Proceedings and Consensus Statement,* Bouchard, C. and Shephard, R. J., Eds., Human Kinetics, Champaign, 1994, 204.
608. Strazzullo, P., Cappuccio, F. P., Trwisan, M., DeLeo, A., Krogh, V., Giorgiom, N., and Mancini, M., Leisure time physical activity and blood pressure in schoolchildren, *Am. J. Epidemiol.,* 127, 726, 1988.
609. Strel, J., *Sports Educational Chart,* Ministry of Education and Sport, Ljubljana, 1997.
610. Strohmayer, H. S., Williams, K., and Schau-George, D., Developmental sequences for catching a small ball: a prelongitudinal screening, *Res. Q. Exercise Sport,* 62, 257, 1991.
611. Stucky-Ropp, R. C. and Di Lorenzo, T. M., Determinants of exercise in children, *Prev. Med.,* 22, 880, 1993.
612. Sungot-Borgen, J. and Larsen, S., Preoccupation with weight and menstrual function in female elite athletes, *Scand. J. Med. Sci. Sports,* 3, 156, 1993.
613. Sunnegardh, J. and Bratteby, L., Maximal oxygen uptake, anthropometry and physical activity in a randomly selected sample of 8- and 13-year-old children in Sweden, *Eur. J. Appl. Physiol.,* 56, 266, 1987.
614. Susanne, C. and Bodszar, E. B., Patterns of secular change of growth and development, in *Secular Growth Changes in Europe,* Bodzsar, E. B. and Susanne, C., Eds., Eötvös University Press, Budapest, 1998, 5.
615. Susanne, C., Bodzsar, E. B., and Castro, S., Factor analysis and somatotyping, are these two physique classification methods comparable?, *Ann. Hum. Biol.,* 25, 405, 1998.

616. Suter, E. and Hawes, M. R., Relationship of physical activity, body fat, diet, and blood lipid profile in youth 10 to 15 years, *Med. Sci. Sports Exercise*, 25, 748, 1993.

617. Szopa, J., Longitudinalna stabilnosc rozwojowa jako metoda okreslania gene- tychnych uwarunkowan rozwoju (Analiza na przykladzie wybranych cech so- matycznych i funkcjonalnuch), *Antropomotoryka*, 5, 35, 1991 (in Polish).

618. Tanner, J. M., *Growth of Adolescence*, 2nd ed., Blackwell Scientific Publications, Oxford, 1962.

619. Tanner, J. M., Whitehouse, R. H., Marshall, W. A., Healy, M. J. R., and Goldstein, H., *Assessment of Skeletal Maturity and Adult Height (TW2 Method)*, Academic Press, New York, 1975.

620. Tanner, J. M., Whitehouse, R. H., Cameron, N., Marshall, W. A., Healy, M. J. R., and Goldstein, H., *Assessment of Skeletal Maturity and Prediction of Adult Height*, 2nd ed., Academic Press, New York, 1983.

621. Taro, O., Jian Ming, Z., Da Jiang, L., and Eiji, K., *The Asian Health Related Fitness Test Results of Chinese Students*, ICHPER 36th World Congress Proceedings, Yokohama, 1993, 98.

622. Telama, R., Laakso, L., and Yang, X., Physical activity and participation in sports of young people in Finland, *Scand. J. Med. Sports*, 4, 65, 1994.

623. Telama, R., Yang, X., Laakso, L., and Viikari, J., Physical activity in childhood and adolescence as a predictor of physical activity in young adulthood, *Am. J. Prev. Med.*, 13, 317, 1997.

624. Tell, G. S. and Vellar, O. D., Physical fitness, physical activity, and cardiovas- cular disease risk factors in adolescents: the Oslo youth study, *Prev. Med.*, 17, 12, 1988.

625. The National Children and Youth Fitness Study II, *JOPERD*, 58, 50, 1987.

626. Thetloff, M., Anthropometric characterization of Estonian girls from 7 to 17 years of age, in *Papers on Anthropology V*, University of Tartu Press, Tartu, 1992, 101.

627. Thomas, J. R. and French, K. E., Gender differences across age in motor per- formance: a meta-analysis, *Psychol. Bull.*, 98, 260, 1985.

628. Thomas, J. R. and Thomas, K. T., Development of gender differences in phys- ical activity, *Quest*, 40, 219, 1988.

629. Thomas, J. R. and Thomas, K. T., What is motor development: where does it belong?, *Quest.*, 41, 203, 1989.

630. Thomas, J. R., Nelson, J. K., and Church, G., A developmental analysis of gender differences in health related physical fitness, *Pediatr. Exercise Sci.*, 3, 28, 1991.

631. Thomas, J. R., Michael, D., and Gallagher, J. D., Effects of training on gender differences in overhand throwing: a brief quantitative literature analysis, *Res. Q. Exercise Sports*, 65, 67, 1994.

632. Thomson, G. H., Weighing for battery reliability and prediction, *Br. J. Physiol.*, 30, 357, 1940.

633. Torun, B., Inaccuracy of applying energy expenditure rates of adults to chil- dren, *Am. J. Nutr.*, 38, 813, 1983.

634. Toselli, S., Geraziani, I., Taraborelli, T., Grispan, A., Tarsitani, G., and Gruppioni, G., Body composition and blood pressure in school children 6 – 14 years of age, *Am. J. Hum. Biol.*, 9, 535, 1997.

635. Trost, S. G., Pate, R. R., Dowda, M., Saunders, R., Ward, D. S., and Felton, G., Gender differences in physical activity and determinants of physical activity in rural fifth grade children, *J. Sch. Health*, 66, 145, 1996.

636. Trost, S. G., Ward, D. S., McGraw, B., and Pate, R. R., Validity of the previous day physical activity recall (PDPAR) in fifth-grade children, *Pediatr. Exercise Sci.*, 11, 341, 1999.

637. Trost, S. G., Pate, R. R., Freedson, P. S., Sallis, J. F., and Taylor, W. C., Using objective physical activity measures with youth: how many days of monitoring are needed? *Med. Sci Sports Exercise*, 32, 426, 2000.

638. Trudeau, F., Laurencelle, L., Tremblay, J., Rajic, M., and Shephard, R. J., Daily primary school physical education: effects on physical activity during adult life, *Med. Sci. Sports Exercise*, 31, 111, 1999.

639. Tucker, L. A. and Hager, R. L., Television viewing and muscular fitness of children, *Percept. Motor Skills*, 82, 1316, 1996.

640. Turek, M., The structure of motoric performance in 6- and 10-year-old children, in *Sport Kinetics '97*, Blaser, B., Ed., Czwalina Verlag, Hamburg, 1997, 208.

641. Twisk, J. W. R., Kemper, H. C. G., and Mellenbergh, G. J., Mathematical and analytical aspects of tracking, *Epidemiol. Rev.*, 16, 165, 1994.

642. Uemura, K. and Piša, Z., Trends in cardiovascular disease mortality industrialized countries since 1930, *World Health Stat. Q.*, 41, 155, 1988.

643. Ulrich, B. D. and Ulrich, D. A., The role of balancing in performance of fundamental motor skills in 3-, 4-, and 5-year-old children, in *Motor Development: Current Selected Research, Vol. 1*, Clark, J. E. and Humphrey, J. H., Eds., Princeton Books, Princeton, 1987.

644. Ulrich, D. A., *Test of Gross Motor Development*, ProEd, Austin, 1985.

645. Updyke, W. F., In search of relevant and credible physical fitness standards for children, *Res. Q. Exercise Sport*, 63, 112, 1992.

646. Updyke, W. and Willett, M. S., Physical fitness trends in American youth 1980-1989, in *Chrysler-AAU Physical Fitness Program*, Bloomington, 1989.

647. USDHHS. *Promotion Physical Activity*, Human Kinetics, Champaign, 1999.

648. Vallerand, R. J. and Reid, G., On the causal effects of perceived competence on intrinsic motivation: a test of cognitive evaluation theory, *J. Sport Psychol.*, 6, 94, 1984.

649. Van Loan, M. D., Assessment of fat-free mass in teenagers: use of TOBEC methodology, *Am. J. Clin. Nutr.*, 52, 586, 1990.

650. Van Loan, M. D. and Mayclin, P. L., Body composition assessment: Dual-energy x-ray absorptiometry (DEXA) compared to reference methods, *Eur. J. Clin. Nutr.*, 46, 125, 1992.

651. Van Praagh, E., Development of anaerobic function during childhood and adolescence, *Pediatr. Exercise Sci.*, 12, 2000, 150.

652. Veldre, G., Somatic Status of 8- to 9-Year-Old Tartu Schoolchildren, M.Sc. dissertation, University of Tartu, Tartu, 1996.

653. Vignerova, J. and Blaha, P., The growth of the Czech child during the past 40 years, in *Secular Growth Changes in Europe*, Bodzsar, B. E. and Susanne, C., Eds., Eötvös University Press, Budapest, 1998, 93.

654. Viru, A., Loko, J., Volver, A., Laaneots, L., Sallo, M., Smirnova, T., and Karelson, K., Alterations in foundations for motor development in children and adolescents, *Coaching Sport Sci. J.*, 1, 11, 1996.

655. Viru, A., Loko, J., Volver, A., Laaneots, L., Karelson, K., and Viru, M., Age periods of accelerated improvement of muscle strength, power, speed and endurance in the age intervals 6 to 18 years, *Biol. Sport*, 15, 211, 1998.
656. Viru, A., Loko, J., Harro, M., Volver, A., Laaneots, L., and Viru, M., Critical periods in the development of performance capacity during childhood and adolescence, *Eur. J. Phys. Educ.*, 4, 75, 1999.
657. Vodak, P. and Wilmore, J., Validity of the 6-minute jog-walk run-walk in estimating endurance capacity in boys, 9 to 12 years of age, *Res. Q. Exercise Sport*, 46, 230, 1975.
658. Volbekiene, V., *Eurofit: Fizinio Pajegumo Testai ir Metodika Lietuvos Kuno [Eurofit: Measurements of Physical Fitness]*, Kulturos in Sporto Departamentas, Vilnius, 1993 (in Lithuanian).
659. Vrijens, J. and Van Cauter, C., Physical performance capacity and specific skills in young soccer players, in *Children and Exercise XI*, Binkhorst, R. A., Kemper, H. C. G., and Saris W. H. M., Eds., Human Kinetics, Champaign, 1988, 285.
660. Wagner, D. R. and Heyward, V. H., Techniques of body composition assessment: a review of laboratory and field methods, *Res. Q. Exercise Sport*, 70, 135, 1999.
661. Waldberg, J. and Ward, D., Role of physical activity in the etiology and treatment of childhood obesity, *Pediatrics*, 2, 82, 1985.
662. Walkley, J., Holland, B., Treloar, R., and Probyn-Smith, H., Fundamental motor skill proficiency of children, *ACHPER National J.*, 40, 11, 1993.
663. Wankell, L. and Pabich, P., The minor sport experience: factors contributing to or detracting from enjoyment, in *Mental Training for Coaches and Athletes*, Orlick, T., Partington, J. T., and Samela, J. H., Eds., Coaching Association in Canada, Ottawa, 1981, 70.
664. Wannamethu, G. and Shaper, A. G., Physical activity and stroke in British middle-aged men, *Br. Med. J.*, 304, 597, 1992.
665. Ward, R., Ross, W. D., Leyland, A. J., and Selbie, S., *The Advanced O-Scale Physique Assessment System*, Kinematrix, Burnaby, 1989.
666. Watkins, J. and Moore, B., The effects of practice on performance in the one mile run test of cardiorespiratory fitness in 12- to 15-year-old girls, *ACHPER Healthy Lifestyle*, 43, 11, 1996.
667. Waxman, M. and Stunkard, A. J., Caloric intake and expenditure of obese boys, *J. Pediatr.*, 96, 187, 1980.
668. Webber, L. S., Cresanta, J. L., Voors, A. W., and Brenson, G. S., Tracking of cardiovascular disease risk factor variables in school-age children. *J. Chronic Dis.*, 36, 647, 1983.
669. Welk, G. J. and Corbin, C. B., The validity of the Tritrac-R3D activity monitor for the assessment of physical activity in children, *Res. Q. Exercise Sport*, 66, 202, 1995.
670. Welk, G. J., Corbin, C. B., and Dale, D., Measurement issues in the assessment of physical activity in children, *Res. Q. Exercise Sport*, 71, 59, 2000.
671. Wessel, J. A., *I. Canadian Physical Education Curriculum*, Hubbard, Northbrook, 1979.
672. Westrate, J. A. and Deurenberg, P., Body composition in children, proposal for a method for calculating body fat skinfold thickness measurements, *Am. J. Clin. Nutr.*, 50, 1104, 1989.

673. Whitehead, J. R., Motivational and Self-perception Outcomes of Physical Fitness on School Children, doctoral dissertation, Arizona State University, Tempe, 1988.
674. Whitehead, J. R., Fitness assessment results — some concepts and analogies, *JOPERD*, 39, 1989.
675. Whitehead, J. R., Pemberton, C. L., and Corbin, C. B., Perspectives on the physical fitness testing of children: the case for a realistic educational approach, *Pediatr. Exercise Sci.*, 2, 111, 1990.
676. Wickstrom, R., *Fundamental Motor Patterns*, 2nd ed., Lea & Febiger, Philadelphia, 1983.
677. Wilczewski, A., Sklad, M., Krawczyk, B., Saczuk, J., and Majle, B., Physical development and fitness of children from urban and rural areas as determined by Eurofit test battery, *Biol. Sport*, 13, 113, 1996.
678. Willee, A. W., *Australian Youth Fitness Survey 1971*, Adelaide, ACHPER, Adelaide, 1973.
679. Williams, D. P., Going, S. B., Lohman, T. G., Harsha, D. W., Srinivasan, S. R., Webber, L. S., and Berenson, G. S., Body fatness and risk for elevated blood pressure, total cholesterol, and serum lipoprotein ratios in children and adolescents, *Am. J. Public Health*, 82, 358, 1992.
680. Williams, K., The temporal structure of the forward roll: inter- and intra-limb coordination, *Hum. Movement Sci.*, 6, 373, 1987.
681. Wilmore, J. H. and McNamara, J. J., Prevalence of coronary heart disease risk factors in boys, 8 to 12 years of age, *J. Pediatr.*, 84, 527, 1974.
682. Winnick, J. R. and Short, F. X., *The Brockport Physical Fitness Test Manual*, Human Kinetics, Champaign, 1999.
683. Wold, B. and Anderssen, H., Health promotion aspects of family and peer influences on sport participation, *Int. J. Sport Pscychol.*, 23, 343, 1992.
684. Wood, T. M. and Safrit, M. J., A comparison of three multivariate models for estimating test battery reliability, *Res. Q. Exercise Sport*, 58, 154, 1987.
685. World Health Organization, Physical Status: the Use and Interpretation of Anthropometry, Report of Expert Committee, WHO, Geneva, 1995.
686. Woynarowska, B., Mukherjee, D., Roche, A. F., and Siervogel, R. M., Blood pressure changes during adolescence and subsequent adult blood pressure level, *Hypertension*, 7, 695, 1985.
687. Wu, Y. T., Nielsen, D. H., Cassady, S. L., Cook, J. S., Janz, K. F., and Hansen, J. R., Cross-validation of bioelectrical impedance analysis of body composition in children and adolescents, *Phys. Ther.*, 73, 320, 1993.
688. Xiaoling Yang, Telama, R., and Leskinen, E., Testing a multidisciplinary model of socialization in physical activity: a 6-year follow-up study, *Eur. J. Phys. Educ.*, 5, 67, 2000.
689. *Young and Active? Policy Framework for Young People and Health-Enhancing Physical Activity*, Health Education Authority, U.K., 1998.
690. Young, C. M., Sipin, S. S., and Roe, D. A., Body composition of pre-adolescent girls, I: density and skinfolds, *J. Am. Diet. Assoc.*, 53, 25, 1968.

Index

O

Obesity, 47, 49, *see also* Body fat
  conclusions and perspectives, 135
  physical activity and, 54, 70–72, 74–75
Overhand throwing studies, 120,
      124–129

P

Pedometers, 61–62
Physical activity, 8–12, 51–76
  assessment of, 52, 55–62
    direct observation, 56, 57
    doubly-layered water technique, 57
    heart rate monitoring, 56, 58–59
    indirect calorimetry, 56, 59
    motion sensors, 56, 59–62, *see also*
      Accelerometers; Pedometers
    questionnaires (self-reports), 56, 58
  basic dimensions of, 52
  biological maturation and, 6
  biological need in children, 52
  body fat/obesity and, 70–72, 74–75
  as compared with exercise, 51, 52
  as compared with physical fitness, 53
  conclusions and perspectives, 132–133
  coronary heart disease (CHD) risk
      and, 11–12
  developmental importance of, 8–9
  family environment and, 7
  gender and, 7
  general considerations, 75–76
  geographical location and, 7–8
  guidelines for, 62–65
    cardiovascular fitness, 63
    comparative criteria, 63–64
    International Consensus
      Conference of Physical Activity
      Guidelines for Adolescents,
      64–65
  health benefits of, 53–55
    disease endpoints in measuring,
      53–54
    risk factors and, 54
  mechanism of protective effect, 10–11
  motor abilities and, 73–74, 85–87
  motor skill development and, 122–123
  national comparisons, 65–68
  physical education and, 9–10
  recent decline in, 9

recommended measurement
    methods, 141
resistance training studies, 72
rural vs. urban environment and, 6–7
self-perception and, 7
sports facilities availability and, 6
tracking studies, 68–69
Physical education, 9–10, 53, 132
  motor skill development and, 121
Physical education teachers, motor
    ability testing recommendations
    for, 91–92
Physical fitness, 79–80, *see also* Motor
    abilities
  as compared with physical activity, 53
  conclusions and perspectives, 133
  health- vs. performance-related, 79–80
  test batteries, 142
Prepubertal vs. pubertal growth, 5
President's Council on Physical Fitness
    and Sports, 96–97
Pubertal growth spurt, 3

Q

Questionnaires (self-reports), in physical
    activity assessment, 56, 58

R

Reliability, of motor ability tests, 92–93
Research directions and recommenda-
    tions, 133–134, 136
Resistance training, 72
Rural versus urban environment, 6–7

S

Secondary sex characteristics, 2
Segmental body impedance, 41–42
Self-perception, 7
Self-reports (questionnaires), in physical
    activity assessment, 57–59
Senegalese physical activities studies,
    67–68
Sex hormones, 2
Sexual dimorphism, *see* Gender
    differences
Sexual maturation, 2–3, 4–5, 48
Singapore NAPFA test battery, 105
Skeletal maturation, 1–2, 4